Natural Computing Series

Thomas Bartz-Beielstein

Experimental Research in Evolutionary Computation

The New Experimentalism

With 66 Figures and 36 Tables

Author

Thomas Bartz-Beielstein
Chair of Algorithm Engineering
and Systems Analysis
Department of Computer Science
University of Dortmund
Otto-Hahn-Str. 14
44227 Dortmund, Germany
thomas.bartz-beielstein@udo.edu

Series Editors

G. Rozenberg (Managing Editor)
rozenber@liacs.nl

Th. Bäck, J.N. Kok, H.P. Spaink
Leiden Center for Natural Computing
Leiden University
Niels Bohrweg 1
2333 CA Leiden, The Netherlands

A.E. Eiben
Vrije Universiteit Amsterdam
The Netherlands

Library of Congress Control Number: 2006922082

ACM Computing Classification (1998): F.1, F.2, G.1.6, G.3, G.4, I.2.8, I.6, J.2

ISSN 1619-7127
ISBN-10 3-540-32026-1 Springer Berlin Heidelberg New York
ISBN-13 978-3-540-32026-5 Springer Berlin Heidelberg New York

Springer is a part of Springer Science+Business Media

springer.com

Printed in Germany

Cover Design: KünkelLopka, Werbeagentur, Heidelberg
Typesetting: by the Author
Production: LE-TEX Jelonek, Schmidt & Vöckler GbR, Leipzig

Printed on acid-free paper 45/3100/YL 5 4 3 2 1 0

To Eva, Leon, Benna, and Zimba

Foreword

Rigorously proven upper and lower run-time bounds for simplified evolutionary algorithms on artificial optimization problems on the one hand and endless tables of benchmark results for real-world algorithms on today's or yesterday's hardware on the other, is that all one can do to justify their invention, existence, or even spreading use? Thomas Bartz-Beielstein gives thoughtful answers to such questions that have bothered him since he joined the team of researchers at the Chair of Systems Analysis within the Department of Computer Science at the University of Dortmund. He brings together recent results from statistics, epistemology of experimentation, and evolutionary computation.

After a long period in which experimentation has been discredited in evolutionary computation, it is regaining importance. This book far exceeds a discussion of often-met points of criticism of the usual experimental approach like missing standards, different measures, and inaccurate and irreproducible results. Also, fundamental objections against the experimental approach are discussed and cleared up. This work shows ways and means to close the gap between theoretical and experimental approaches in algorithm engineering. It becomes clear that statistical tests are the beginning and not the end of experimental analyses. Vital in this context is the differentiation between statistically relevant and scientifically meaningful results, which is clearly developed by Thomas Bartz-Beielstein.

The results of this book—especially the sequential parameter optimization developed in Chap. 7—can directly be applied. They have been used in the evolutionary optimization of algorithmic chemistries, in chemical engineering, machining technology, electrical engineering, and for other real-world problems such as the optimization of elevator group controllers.

The impact of the author's insights goes beyond the field of computer science. The techniques presented are also of great interest for designing procedures in numerical mathematics.

I would like to call this book a first innovative attempt—I do not know any other—to create a theory of trying. Impressive is the wide epistemological arc the author draws from the philosophy of science over the behavioral sciences to numerical mathematics and computer science to legitimate a method that is commonly applied by optimization practitioners. He lays a solid base for scientific experimentation in computer science and proposes a course of action that is reliable as far as possible.

However, experiments require a lot of work, so the reader may be warned: Performing a good experiment is as demanding as proving a new theorem.

Dortmund, November 2005 *Hans-Paul Schwefel*

Preface

Before we go into medias res, I would like to acknowledge the support of many people who made this book possible.

First and foremost, I would like to thank Hans-Paul Schwefel, the head of the Chair of Systems Analysis, for providing a cooperative and stimulating work atmosphere. His thoughtful guidance and constant support in my research were very valuable and encouraging. This book is based on my dissertation "New Experimentalism Applied to Evolutionary Computation" (Bartz-Beielstein 2005b). I am thankful to Peter Buchholz for his kindness in being my second advisor, and I would like thank Ingo Wegener for valuable discussions.

Thomas Bäck supported my scientific research for many years, beginning when I was a student and working at the Chair of Systems Analysis and during the time I did work for NuTech Solutions. He also established the contact to Sandor Markon, which resulted in an inspiring collaboration devoted to questions related to elevator group control and the concept of threshold selection. Sandor Markon also provided guidance in Korea and Japan, which made my time there very enjoyable.

I greatly appreciated the discussions with Dirk Arnold relating to threshold selection. They built the cornerstone for a productive research that is devoted to selection and decision making under uncertainty.

The first official presentation of the ideas from this book during the CEC tutorial on experimental research in evolutionary computation in 2004 was based on the collaboration and the helpful discussions with Mike Preuß. Tom English's support during the preparation and presentation of this tutorial were very comforting. I also very much enjoyed the constructive exchange of information with the people from the evolutionary computation "task force," Steffen Christensen, Gwenn Volkert, and Mark Wineberg. Many thanks go to Jürgen Branke for inspiring discussions about experimental approaches and to Burkhard Hehenkamp and Thomas Stützle for their comments on early versions of this work.

My colleagues Boris Naujoks, Karlheinz Schmitt, and Christian Lasarczyk shared their knowledge and resources, helped in many discussions to clarify my ideas, and made the joint work a very fortunate experience. Konstantinos E. Parsopoulos and Michael N. Vrahatis aroused my interest in particle swarm optimization. Discussions with students, especially with Christian Feist, Marcel de Vegt, and Daniel Blum, have been a valuable source of inspiration during this research.

This book would not have been completed without the help from Ronan Nugent, who supported the editorial process.

Additional material (exercises, solutions to selected exercises, program sources) is available under the following link:
`http://www.springer.com/3-540-32026-1`

Dortmund, November 2005 *Thomas Bartz-Beielstein*

Contents

Part I

Basics

1

Research in Evolutionary Computation

This work tries to lay the groundwork for experimental research in evolutionary computation. We claim that experiments are necessary—a purely theoretical approach cannot be seen as a reasonable alternative. Our approach is related to the discipline of *experimental algorithmics*, which provides methods to improve the quality of experimental research. However, many approaches from experimental algorithmics are based on Popperian paradigms:

1. No experiment without theory.
2. Theories should be falsifiable.

Following Hacking (1983) and Mayo (1996), we argue that:

1*. An experiment can have a life of its own.
2*. Falsifiability should be complemented with verifiability.

This concept, known as the *new experimentalism*, is an influential discipline in the modern philosophy of science. It provides a statistical methodology to learn from experiments. For a correct interpretation of experimental results, it is crucial to distinguish the statistical significance of an experimental result from its scientific meaning. This work attempts to introduce the concept of the new experimentalism in evolutionary computation.

1.1 Research Problems

At present, it is intensely discussed which type of experimental research methodologies should be used to improve the acceptance and quality of *evolutionary algorithms* (EA). A broad spectrum of presentation techniques makes new results in *evolutionary computation* (EC) almost incomparable. Sentences like "This experiment was repeated ten times to obtain significant results" or "We have proven that algorithm A is better than algorithm B" can still be found in current EC publications. Eiben & Jelasity (2002) explicitly list four problems:

Problem 1.1. The lack of standardized test-functions, or benchmark problems.

Problem 1.2. The usage of different performance measures.

Problem 1.3. The impreciseness of results, and therefore no clearly specified conclusions.

Problem 1.4. The lack of reproducibility of experiments.

These problems provide guidelines for our analysis and will be reconsidered in Chap. 9. In fact, there is a gap between theory and experiment in evolutionary computation. How to promote good standards and quality of research in the field of evolutionary computation was discussed during the *Genetic and Evolutionary Computation Conference* (GECCO) in 2002. Bentley noted:

> Computer science is dominated by the need to publish, publish, publish, but sometimes this can happen at the expense of research. All too often poor papers, clumsy presentations, bad reviews or even bad science can clutter a conference, causing distractions from the more carefully prepared work (Bentley 2002).

There is a great demand for these topics, as one can see from the interest in tutorials devoted to these questions during two major conferences in evolutionary computation, the *Congress on Evolutionary Computation* (CEC) and GECCO (Bartz-Beielstein et al. 2003d; Wineberg & Christensen 2004; Bartz-Beielstein & Preuß 2004, 2005a, b).

1.2 Background

Evolutionary computation shares these problems with other scientific disciplines such as simulation, artificial intelligence, numerical analysis, or industrial optimization (Dolan & More 2002). Cohen's survey of 150 publications from the proceedings of the Eighth National Conference on Artificial Intelligence, which was organized by the *American Association for Artificial Intelligence*, "gave no evidence that the work they described has been tried out on more than a single example problem" (Cohen et al. 2000). He clearly demonstrated that there is no essential synergy between experiment and theory in these papers.

Cohen (1995) not only reported these negative results, he also provided valuable examples for how empirical research can be related to theory. Solutions from other disciplines that have been applied successfully for many years might be transferable to evolutionary computation. We have chosen four criteria to classify existing experimental research methodologies that have a lot in common with our approach. First, we can mention effective approaches.

They find a solution but are not very efficient and are not focused on understanding. Greedy, or brute-force approaches belong to this group. Second, meta-algorithms can be mentioned. They might locate good parameter sets, though without providing much insight into how sensitive performance is to parameter changes. Third, approaches that model problems of mostly academic interest can be listed. These approaches consider artificial test functions or infinite population sizes. Finally, the fourth category comprehends approaches that might be applicable to our problems although they have been developed with a different goal. Methods for deterministic computer experiments can be mentioned here. We will give a brief overview of literature on experimental approaches from these four domains.

1.2.1 Effective Approaches

The methodology presented in this book has its origins in statistical *design of experiments* (DOE). But classical DOE techniques as used in agricultural or industrial optimization must be adapted if applied to optimization models since stochastic optimization uses pseudorandom numbers (Fisher 1935). Randomness is replaced by pseudorandomness. For example, blocking and randomization, which are important techniques to reduce the systematic influence of different experimental conditions, are unnecessary in computer-based optimization. The random number seed is the only random element during the optimization run.

Classical DOE techniques are commonly used in simulation studies—a whole chapter in a broadly cited textbook on simulation describes experimental designs (Law & Kelton 2000). Kleijnen (1987, 1997) demonstrated how to apply DOE in simulation. As simulation is related to optimization (simulation models equipped with an objective function define a related optimization problem), we suggest the use of DOE for the analysis of optimization problems and algorithms (Kelton 2000).

This work is not the first attempt to use classical DOE methods in EC. However, our approach takes the underlying problem instance into account. Therefore, we do not try to draw any problem-independent conclusions such as: "The optimal mutation rate in genetic algorithms is 0.1." In addition, we propose an approach that is applicable if a small amount of function evaluations are available only. Schaffer et al. (1989) proposed a complete factorial design experiment that required 8400 run configurations; each configuration was run to 10,000 fitness function evaluations. Feldt & Nordin (2000) use statistical techniques for designing and analyzing experiments to evaluate the individual and combined effects of genetic programming parameters. Three binary classification problems are investigated in a total of 7 experiments consisting of 1108 runs of a machine code genetic programming system. Myers & Hancock (2001) present an empirical modeling of genetic algorithms. This approach requires 129,600 program runs. François & Lavergne (2001) demonstrate the applicability of *generalized linear models* (GLMs) to design evolutionary algorithms. Again, data sets of size 1000 or even more are necessary, although a simplified evolutionary algorithm with 2 parameters only is designed.

As we include methods from computational statistics, our approach can be seen as an extension of these classical approaches. Furthermore, classical DOE approaches rely strongly on hypothesis testing. The reconsideration of the framework of statistical hypothesis testing is an important aspect in our approach.

1.2.2 Meta-Algorithms

The search for useful parameter settings of algorithms itself is an optimization problem. Optimization algorithms, so called meta-algorithms, can be defined to accomplish this task. Meta-algorithms for evolutionary algorithms have been proposed by many authors (Bäck 1996; Kursawe 1999). But this approach does not solve the original problem completely, because it requires the determination of a parameter setting of the meta-algorithm.

Additionally, we argue that the experimenter's skill plays an important role in this analysis. It cannot be replaced by automatic rules. The difference between automatic rules and learning tools is an important topic discussed in the remainder of this book.

1.2.3 Academic Approaches

Experimental algorithmics offer methodologies for the design, implementation, and performance analysis of computer programs for solving algorithmic problems (Demetrescu & Italiano 2000; Moret 2002). McGeoch (1986) examined the application of experimental, statistical, and data analysis tools to problems in algorithm analysis. Barr & Hickman (1993) and Hooker (1996) tackled the question how to design computational experiments and how to test heuristics. Aho et al. (1997) tried "to achieve a greater synergy between theory and practice."

Most of these studies were focused on *algorithms*, and not on *programs*. Algorithms can be analyzed on a sheet of paper, whereas the analysis of programs requires real hardware. The latter analysis includes the influence of rounding errors or limited memory capacities. We will use both terms simultaneously, because whether we refer to the algorithm or the program will be clear from the context.

Compared to these goals, our aim is to provide methods for very complex real-world problems, when only a few optimization runs are possible, i.e., optimization via simulation. The elevator supervisory group controller study discussed in Beielstein et al. (2003a) required more than a full week of round-the-clock computing in a batch job processing system to test 80 configurations.

Our methods are applied to real computer programs and not to abstract algorithms. A central topic in complexity theory is to answer the question NP

$\neq$ P. It is assumed that the class of problems that can be solved *nondeterministically in polynomial time* (NP) is different from the class of problems that can be solved in *polynomial time* (P). Problems in NP are—in contrast to problems in P—considered difficult and not efficiently solvable. However, analyses from complexity theory are not sufficient for some problems (Weihe et al. 1999). Many simple problems belong to NP. Niedermeier (2003) develops a recent approach to overcome this "dilemma of NP-hardness." Furthermore, there is an interesting link between programs (experimental approach) and algorithms (complexity theory) as discussed in Example 1.1.

Example 1.1 (Hooker 1994). Consider a small subset of very special *traveling salesperson problems* (TSP) T. This subset is NP-complete, and any class of problems in NP that contains T is ipso facto NP-complete. Consider the class P' that consists of all problems in P and T. As P' contains all easy problems in the world, it seems odd to say that problems in P' are hard. But P' is no less NP-complete than TSP. Why do we state that TSP is hard? Hooker (1994) suggests that "we regard TSP as a hard class because *we in fact find problems in TSP to be hard in practice*." We acknowledge that TSP contains many easy problems, but we are able to generate larger and larger problems that become more and more difficult. Hooker suggests that it is this empirical fact that justifies our saying that TSP contains characteristically hard problems. And, in contrast to P', TSP is a natural problem class, or as philosophers of science would say, a natural kind. ■

1.2.4 Approaches with Different Goals

Although our methodology has its origin in DOE, classical DOE techniques used in agricultural and industrial simulation and optimization tackle different problems and have different goals.

Parameter control deals with parameter values (*endogenous strategy parameters*) that are changed during the optimization run (Eiben et al. 1999). This differs from our approach, which is based on parameter values that are specified before the run is performed (*exogenous strategy parameters*). The assumption that specific problems require specific EA parameter settings is common to both approaches.

Design and analysis of computer experiments (DACE) as introduced in Sacks et al. (1989) models the deterministic output of a computer experiment as the realization of a stochastic process. The DACE approach focuses entirely on the correlation structure of the errors and makes simplistic assumptions about the regressors. It describes "how the function behaves," whereas regression as used in classical DOE describes "what the function is" (Jones et al. 1998, p. 14). DACE requires other experimental designs than classical DOE, e.g., Latin hypercube designs (McKay et al. 1979). We will discuss differences and similarities of these designs and present a methodology for how DACE can be applied to stochastic optimization algorithms.

Despite the differences mentioned above, it might be beneficial to adapt some of these well-established ideas from other fields of research to improve the acceptance and quality of evolutionary algorithms.

1.3 Common Grounds: Optimization Runs Treated as Experiments

Gregory et al. (1996) performed an interesting study of dynamic scheduling that demonstrates how synergetic effects between experiment and theory can evolve. Johnson et al. (1989, 1991) are seminal studies of simulated annealing. Rardin & Uzsoy (2001) presented a tutorial that discusses the experimental evaluation of heuristic search algorithms when the complexities of the problem do not allow exact solutions. Their tutorial described how to design test instances, how to measure performance, and how to analyze and present the experimental results. They demonstrated pitfalls of commonly used measures such as the algorithm-to-optimal ratio, that measures how close an algorithm comes to producing an optimal solution.

Birattari et al. (2002) developed a "racing algorithm" for configuring metaheuristics that combines blocking designs, nonparametric hypothesis testing, and Monte Carlo methods. The aim of their work was "to define an automatic hands-off procedure for finding a good configuration through statistical guided experimental evaluations." This is unlike the approach presented here, which provides means for understanding algorithms' performance (we will use datascopes similar to microscopes in biology and telescopes in astronomy). However, Chiarandini et al. (2003) demonstrate that racing can be used interactively and not only as a monolithic block. These studies—although based on classical DOE techniques only—are closely related to our approach.

Optimization runs will be treated as experiments. In our approach, an experiment consists of a problem, an environment, an objective function, an algorithm, a quality criterion, and an initial experimental design. We will use methods from computational statistics to improve, compare, and understand algorithms' performances. The focus in this work lies on natural problem classes: Its elements are problems that are based on real-world optimization problems in contrast to artificial problem classes (Eiben & Jelasity 2002). Hence, the approach presented here might be interesting for optimization practitioners who are confronted with a complex real-world optimization problem in a situation where only few preliminary investigations are possible to find good parameter settings.

Furthermore, the methodology presented in this book is applicable a priori to tune different parameter settings of two algorithms to provide a fair comparison. Additionally, these methods can be used in other contexts to improve the optimization runs. They are applicable to generate systematically feasible starting points that are better than randomly generated initial points, or to guide the optimization process to promising regions of the search space.

Meta-model assisted search strategies as proposed in Emmerich et al. (2002) can be mentioned in this context. Jin (2003) gives a survey of approximation methods in EC.

Before introducing our understanding of experimental research in EC, we may ask about the importance of experiments in other scientific disciplines. For example, the role of experiments in economics changed radically during recent decades.

1.3.1 Wind Tunnels

The path-breaking work of Vernon L. Smith (2002 Nobel Prize in Economics together with Daniel Kahneman) in experimental economics provided criteria to find out whether economic theories hold up in reality. Smith demonstrated that a few relatively uninformed people can create an efficient market. This result did not square with theory. Economic theory claimed that one needed a horde of "perfectly informed economic agents." He reasoned that economic theories could be tested in an experimental setting: an economic wind tunnel. Smith had a difficult time getting the corresponding article published (Smith 1962). Nowadays this article is regarded as the landmark publication in experimental economics.

Today, many cases of economic engineering are of this sort. Guala (2003) reports that before "being exported to the real world" the auctions for mobile phones were designed and tested in the economic laboratory at Caltech. This course of action suggests that experiments in economics serve the same function that a wind tunnel does in aeronautical engineering. But, the relationship between the object of experimentation and the experimental tool is of importance: How much reductionism is necessary to use a tool for an object? Table 1.1 lists some combinations. Obviously some combinations fit very well, whereas others make no sense at all.

Table 1.1. Relationship between experimental objects and experimental tools. Some combinations, for example, reality–computer, require some kind of reductionism. Others, for example, algorithm–wind tunnel, are useless

Object of experimentation	Experimental tool
Reality	Computer
Reality	Thought experiment
Reality	Wind tunnel
Airplane	Computer
Airplane	Thought experiment
Airplane	Wind tunnel
Algorithm	Computer
Algorithm	Thought experiment
Algorithm	Wind tunnel

We propose an experimental approach to analyze algorithms that is suitable to discover important parameters and to detect superfluous features. But before we can draw conclusions from experiments, we have to take care that the experimental results are correct. We have to provide means to control the error, because we cannot ensure that our results are always sound. Therefore the concept of the new experimentalism is regarded next.

1.3.2 The New Experimentalism

The new experimentalism is an influential trend in recent philosophy of science that provides statistical methods to set up experiments, to test algorithms, and to learn from the resulting errors and successes. The new experimentalists are seeking a relatively secure basis for science, not in theory or observation but in experiment. To get the apparatus working for simulation studies is an active task. Sometimes the recognition of an oddity leads to new knowledge. Important representatives of the new experimentalism are Hacking (1983), Galison (1987), Gooding et al. (1989), Mayo (1996), and Franklin (2003). Deborah Mayo, whose work is in the epistemology of science and the philosophy of statistical inference, proposes a detailed way in which scientific claims are validated by experiment. A scientific claim can only be said to be supported by experiment if it passes a severe test. A claim would be unlikely to pass a severe test if it were false. Mayo developed methods to set up experiments that enable the experimenter, who has a detailed knowledge of the effects at work, to learn from error.

1.4 Overview of the Remaining Chapters

The first part of this book (Chaps. 1 to 6) develops a solid statistical methodology, which we consider to be essential in performing computer experiments. The second part, which is entitled "Results and Perspectives" (Chaps. 7 and 8) describes applications of this methodology.

New concepts for an objective interpretation of experimental results are introduced. Each of the following seven chapters closes with a summary of the key points. The concept of the new experimentalism for computer experiments and central elements of an understanding of science are discussed in Chap. 2. It details the difference between demonstrating and understanding, and between significant and meaningful. To incorporate these differences, separate models are defined: models of hypotheses, models of experimental tests, and models of data. This leads to a reinterpretation of the *Neyman–Pearson theory of testing* (NPT). Since hypothesis testing can be interpreted objectively, tests can be considered as learning tools. Analyzing the frequency relation between the acceptance (and the rejection) of the null hypothesis and the difference in means enables the experimenter to learn from errors. This concept of learning tools provides means to extend Popper's widely accepted claim that theories should be falsifiable.

Statistical definitions for Monte Carlo methods, classical design and analysis of experiments, tree-based regression methods, and modern design and analysis of computer experiments techniques are given in Chap. 3. A bootstrap approach that enables the application of learning tools if the sampling distribution is unknown is introduced. This chapter is rather technical, because it summarizes the relevant mathematical formulas.

Computer experiments are conducted to improve and to understand the algorithm's performance. Chapter 4 presents optimization problems from evolutionary computation that can be used to measure this performance. Before an elevator group control problem is introduced as a model of a typical real-world optimization problem, some commonly used test functions are presented. Problems related to test suites are discussed as well.

Different approaches to set up experiments are discussed in Chap. 5. Classical and modern designs for computer experiments are introduced. A sequential design based on DACE that maximizes the expected improvement is proposed.

Search algorithms are presented in Chap. 6. Classical search techniques, for example, the Nelder–Mead "simplex" algorithm, are presented as are stochastic search algorithms. The focus lies on particle swarm optimization algorithms, which build a special class of bioinspired algorithms.

The discussion of the concept of optimization provides the foundation to define performance measures for algorithms in Chap. 7. A suitable measure reflects requirements of the optimization scenario or the experimental environment. The measures are categorized with respect to effectivity and efficiency. Now, the necessary components according to the discussion in the previous chapters to perform computer experiments are available: a problem, an environment, an objective function, an algorithm, a quality criterion, and an experimental design. After summarizing a classical DOE approach of finding better suited exogenous parameters (tuning), a sequential approach that comprehends methods from computational statistics is presented. To demonstrate that our approach can be applied to any arbitrary optimization algorithm, several variants of optimization algorithms are tuned. Tools from error statistics are used to decide whether statistically significant results are scientifically meaningful.

Chapter 8 closes the circle opened in Chap. 2 on the discussion of testing as an automatic rule and as a learning tool. Provided with the background from Chap. 2, the aim of Chap. 8 is to propose a method to learn from computer experiments and to understand how algorithms work. Various schemes for selection under noise for direct search algorithms are presented. Threshold selection is related to hypothesis testing. It serves as an example to clarify the difference between tests as rules of inductive behavior and tests as learning tools. A summary and an outlook conclude this book in Chap. 9.

Introducing the new experimentalism in evolutionary computation provides tools for the experimenter to understand algorithms and their interactions with optimization problems. Experimentation is understood as a means for testing hypotheses, the experimenter can learn from error and control the consequences of his decisions. The methodology presented here is based on the statistical methods most widely used by today's practicing scientists. It might be able "to offer genuine hope for a recovery of some of the solid intuitions of the past about the objectivity of science"(Ackermann 1989).

2

The New Experimentalism

> The physicist George Darwin used to say that every once in a while one should do a completely crazy experiment, like blowing the trumpet to the tulips every morning for a month. Probably nothing would happen, but what if it did?
>
> —Ian Hacking

In this chapter we discuss the role of experiments in evolutionary computation. First, problems related to experiments are presented. Objections stated by theoreticians, for example, "Algorithms are formal objects and should be treated formally," are discussed. After considering these objections, we present an experimental approach in evolutionary computation. Important goals for scientific research in evolutionary computation are proposed. Experimental algorithmics is an influential discipline that provides widely accepted methods to tackle these scientific goals. It is based on a Popperian understanding of science. After introducing the concept of models in science, the new experimentalism is presented. It goes beyond Popper's concept that only results that are falsifiable should be treated as scientific. The new experimentalists claim that the experimenter can learn from experiments. Mayo introduced an approach based on the Neyman–Pearson theory of statistical testing that enables the experimenter to perform experiments in an objective manner. It is important to note that statistically significant results are not automatically meaningful. Therefore some space must be left between the statistical result and its scientific import. Finally, the relationship between theory and practice is reconsidered.

2.1 The Gap Between Demonstrating and Understanding

We first start with a comparison of two parameterizations of a stochastic global optimization method. This comparison is based on real optimization data, but it is kept as simple as possible for didactical purposes.

Example 2.1 (PSO swarm size). Analyzing a *particle swarm optimization algorithm* (PSO), we are interested in testing whether or not the swarm size has a significant influence on the performance of the algorithm. The 10-dimensional Rosenbrock function was chosen as a test function. Based on the parameterization in Shi & Eberhart (1999), the swarm sizes were set to 20 and 40. The corresponding settings will be referred to as run PSO(20) and PSO(40), respectively. The question is whether the increased swarm size improves the performance of the PSO. As in Shi & Eberhart (1999), a random sample is drawn from each of the two populations. The average performance $\overline{y}_1$ of $n = 50$ runs of PSO(20) is 108.02, whereas the average performance $\overline{y}_2$ of $n = 50$ runs of PSO(40) is 56.29. The same number of function evaluations was used in both settings. The number of runs n is referred to as the *sample size*, and $\overline{y}$ denotes the *sample mean*. ■

Example 2.1 demonstrates at first sight, that the hypothesis (H)

(H-2.1) PSO(40) outperforms PSO(20)

is correct. But can we really be sure that (H-2.1) is true? Is this result statistically significant? Are the influences of other parameters on the algorithm's performance negligible? Are 50 repeats sufficient? How does the run length, that is, the number of iterations, influence the result? However, even if we assume that (H-2.1) is correct, what can we learn from this conclusion?

As will be demonstrated later on and as some readers already know, choosing a suitable parameterization enables the experimenter to demonstrate anything—algorithm A is better than algorithm B, or the other way round. The remainder of this book deals with questions related to these problems and provides a methodology to perform comparisons in a statistically sound manner.

2.1.1 Why Do We Need Experiments in Computer Science?

It is a difficult task to set up experiments correctly. Experimental results may be misleading. So one may ask why to perform computer experiments at all.

There are theoretical and empirical approaches to study the performance of algorithms. In contrast to some researchers who consider merely the former as scientific, many practitioners are convinced that the theoretical approach alone is not well-suited to judge an algorithm's performance.

Why is the empirical work viewed as unscientific? One reason might be the lack of standards for empirical methods. Empirical work is sometimes considered as "lowbrow or unsophisticated" (Hooker 1994). Additionally, the irreproducibility of the results discredits empirical approaches (Eiben & Jelasity 2002). But these are problems that can be mastered, at least in principle. The main objection against empirical work lies deeper. Hooker hits the nail on the head with the following characterization: The main objection against empirical work is comparable to the uneasiness that arises when "verifying that opposite interior angles are equal by measuring them with a protractor" (Hooker 1994). This can be formulated as:

Statement 2.1. Algorithms are defined as formal systems and should be studied with formal methods.

Reasoning that many founders of modern science like Galilei, Descartes, Leibniz, or Newton studied formal systems with empirical methods does not give a completely satisfactory response to this objection. After discussing the role of models in science, we will reconsider this objection and give a well-founded answer that uses some fundamental concepts from the philosophy of science. However, even under the assumption that Statement 2.1 is true, studying formal systems is not as trivial as it might appear at first sight. Severe objections arise when Statement 2.1 is considered in detail. The construction of a self-explanatory formal system requires a huge complexity. These systems cannot be applied in real-life situations (Mertens 1990). This may be one reason for the failure of the enthusiastically propagated set-theoretical approach to mathematics in primary schools in the 1960s. Nowadays it is widely accepted that the Peano axioms do not provide a suitable context to introduce the system of whole numbers for primary schools (Athen & Bruhn 1980).

But not only its complexity makes the formal approach difficult. As Statement 2.1 cannot be proven, it is rather subjective. It is obvious that:

1. Algorithms treated as formal systems require some kind of reductionism.
2. Reductionism works in some cases, but fails in others.

Based on our subjective experience as experimenters, we can claim that:

Statement 2.2. Reductionism often fails in algorithmic science.

Hooker gives an example that emphasizes the importance of the right level of reductionism or abstraction that provides an understanding of the underlying phenomena: Investigating the behavior of algorithms with formal methods is like applying quantum physics to geology to understand plate tectonics. Even if one can in principle deduce what the algorithms are going to do, we would not understand why they behave as they do (Hooker 1994). As will be seen in the following, the concept of model plays a central role in tackling these problems.

Comparing mathematical models and experiments, the following statements are true:

1. Results from mathematical models are *more* certain than results from experiments.
2. Results from mathematical models are *less* certain than results from experiments.

As the conclusions must follow from the premises, mathematical models are more certain. This justifies the first statement. However, as these premises are more hypothetical and arbitrary, the conclusions are less certain. This confirms the second statement. Both mathematical models and experiments deliver only hypotheses. Or, as stated by Einstein: "As far as the laws of mathematics refer to reality, they are not certain; and as far as they are certain, they do not refer to reality" (Newman 1956).

Summarizing, one can claim that a solely theoretically oriented approach is not completely satisfactory. We would like to mention a few more reasons why experiments are useful:

- Theories may be incomplete; they may have holes. Consider the *Nelder–Mead simplex algorithm* (Nelder & Mead 1965), one of the most popular methods for nonlinear unconstrained optimization. "At present there is no function in any dimension greater than 1 for which the original Nelder–Mead algorithm has been proved to converge to a minimizer" (Lagarias et al. 1998).
- Observations and experiments may suggest new theories. Existing theories can be tested and refined by experiments. The 2004 NASA gravity probe B mission can be mentioned here. Gravity probe B is an experiment being developed by NASA and Stanford University to test two extraordinary, unverified predictions of Einstein's general theory of relativity.
- Experiments can bridge the gap between theory and practice: Experiences from the collaborative research center "Design and Management of Complex Technical Processes and Systems by Means of Computational Intelligence Methods" in Dortmund show that the engineer's view of complexity differs from the theoretician's view (Schwefel et al. 2003).
- Digital computers use a finite number of decimal places to store data. To check theoretically derived convergence results, computer experiments that consider rounding errors, have to be performed (Schwefel 1995).
- Worst case $\neq$ average case: As worst-case scenarios can theoretically be analyzed more easily than average-case scenarios, many theoretical results are related to the former. But the real world is "mostly" an average-case scenario (Briest et al. 2004).
- To obtain average-case results, it is necessary to define a probability distribution over randomly generated problem instances. But, this distribution is "typically unreflective of reality" (Hooker 1994). Real-world optimization problems are not totally random; they possess some kind of structure. To model this structure, Kan (1976) introduced job-correlated processing times for scheduling problems. As a consequence, theoretical lower-bound calculations that perform very well for randomly generated problem instances provide "an extremely poor bound on job-correlated problems" (Whitley et al. 2002). We will come back to this problem in Chap. 4.
- And last, but not least: What is c in $O(n) + c$?

Many theoreticians would accept the point of view that experiments are useful—at least when they support theories.

If experiments are elements of science—this point has not been clarified yet—an experimental methodology to deal with important research questions is needed. Over ten years ago, Hooker (1994), coming from the operations research community, postulated to "build an empirical science of algorithms." The key ingredients of his empirical science are statistical methods and empirically based theories that can be submitted to rigorous testing. We claim that the time is ripe to transfer these ideas to the field of evolutionary algorithms. First, we will ask, "What is the role of science?" Or, to be more concrete, "Which are important research topics especially in evolutionary computation?"

2.1.2 Important Research Questions

We claim that important elements of research in evolutionary computation comprise tasks like:

Goal 2.1 (Discovery). Specifying optimization problems and analyzing algorithms. Which are important parameters; what should be optimized; what happens if new operators are implemented?

Goal 2.2 (Comparison). Comparing the performance of competing search heuristics such as evolutionary algorithms, simulated annealing, and particle swarm optimization, etc.

Goal 2.3 (Conjecture). It might be good to demonstrate performance, but it is better to explain performance. Understanding and further research, based on statistics and visualization techniques, play important roles.

Goal 2.4 (Quality). Improving the robustness of the results obtained in optimization or simulation runs. Robustness includes insensitivity to exogenous factors that can affect the algorithms' performance and minimization of the variability around the solutions obtained (Montgomery 2001).

These four *research goals* (RG) are also discussed in the discipline of experimental algorithmics. As this discipline has gained much attention in recent years, we will present this approach to establish experiments as scientific means first. (Pre)experimental planning is of importance in experimental algorithmics.

2.2 Experimental Algorithmics

2.2.1 Preexperimental Planning

Preexperimental planning has a long tradition in other scientific disciplines. For instance, Coleman & Montgomery (1993) present a checklist for the preexperimental planning phases of an industrial experiment. It covers the following topics: objectives of the experiment, a performance measure, relevant background on response and control variables, a list of response variables and control variables, a list of factors to be "held constant", known interactions, proposed analysis techniques, etc. The differences between analytical and empirical studies are discussed in Anderson (1997). Good empirical work must pass the following tests: "It must be both convincing and interesting" (Anderson 1997). Moret (2002) gives a suitable characterization of "interesting": "Always look beyond the obvious measures!" In this context, we recommend including factors that should have no effect on the response, such as the random seed in the model. This is one example for "blowing the trumpet to the tulips."

2.2.2 Guidelines from Experimental Algorithmics

As we have classified important parameters of the algorithm to be analyzed, and have defined a measure for its performance, we can conduct experiments to assess the significance of single parameters such as population size or selective pressure. Optimization runs are treated as experiments. We begin by formulating a hypothesis, then we set up experiments to gather data that either verify or falsify this hypothesis. *Guidelines* (GL) from experimental algorithmics to set up and to perform experiments read as follows (Moret 2002):

Guideline 2.1 (Question). State a clear set of objectives. Formulate a question or a hypothesis. Typical questions or hypotheses read: "Is the selective pressure $\nu = 5$ a good choice for the optimization problem under consideration?", or "PSO works better when the swarm size is ten times the dimension of the search space compared to a parameterization that uses a fixed swarm size."

Guideline 2.2 (Data collection). After an experimental design is selected, simply gather data.

Guideline 2.3 (No peeking). Do not modify the hypothesis until all data have been collected. This guideline is also known in the philosophy of statistics as the *no peeking* rule.

Guideline 2.4 (Analysis). Analyze the data to test the hypothesis stated above.

Guideline 2.5 (Next Cycle). In the next cycle of experimentation a new hypothesis can be tested, i.e. "$\nu = 5$ is a good choice, because..."

This procedure complies with Popper's position that "knowledge results when we accept statements describing experience that contradict and hence refute our hypotheses; thus a deductive rather than an inductive relation holds between theoretical knowledge and experience. Experience teaches us by correcting our errors. Only hypotheses falsifiable by experience should count as scientific" (Jarvie 1998). Or, as Sanders introduces the related discipline of algorithm engineering: "Algorithm engineering views design, analysis, implementation, and experimental analysis of algorithms as an integrated process in the tradition of Popper's scientific method" (Sanders 2004). Figure 2.1 depicts

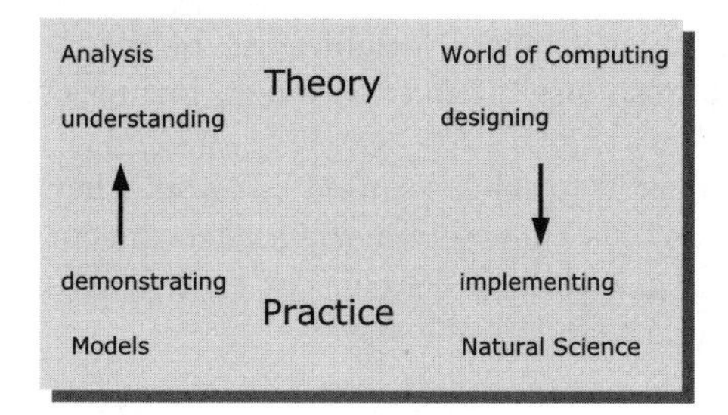

Fig. 2.1. A first approach to model the relationship between theory and practice. Practice can benefit from theory, and vice versa. Demonstrating good results is only the first step in the scientific process, whereas nature's reality can be seen as the judge of a scientific theory (Bartz-Beielstein 2003)

a commonly accepted view on the relationship between theory and experiment. This position is nowadays broadly accepted in the computer science community. However, it is intensely discussed in the philosophy of science. Results from these discussions have a direct impact on the experimental methodology in evolutionary computation. Therefore we will present the fundamental ideas in the following. After introducing the framework of the new experimentalists in Sect. 2.5, Popper's position will be reconsidered.

To introduce the concept of models as a central element of science, we describe an inherent problem of nearly any real-world situation: noise.

2.3 Observational Data and Noise

Even if a scientific hypothesis or claim describes a phenomenon of investigation correctly, observational data may not precisely agree with it. The accuracy and precision of such data may be limited by measurement errors or inherent fluctuations of the response quantity, for example turbulences. Another source of distortion may lie in the inherent probabilistic nature of the scientific hypotheses. Moreover, the observational data are discrete in contrast to scientific hypotheses that may refer to continuous values (Mayo 1983).

Although computer programs are executed deterministically, evolutionary computation has to cope with noise. Stochastic optimization uses pseudorandom numbers. Randomness is replaced by pseudorandomness. As common or antithetic seeds can be used, the optimization practitioner has much more control over the noise in the experiments and can control the source of variability (Kleijnen 1997). The different optimization runs for one specific factor combination can be performed under exactly the same conditions—at least in principle: Even under exactly the same conditions different hardware can produce unexpected results. To compare different run configurations under similar conditions *variance-reduction techniques* (VRT) such as *common random numbers* (CRN) and antithetic variates can be applied (Law & Kelton 2000).

Random error or noise can be classified as an unsystematic effect. Systematic errors, for example the selection of a wrong regression model, are referred to as *bias*. Santner et al. (2003) distinguish control variables, noise (or environmental) variables, and model variables. Control variables can be set by the experimenter to control the output (response) of an experiment, but noise variables depend on the environment.

Statistical methods can be used to master problems caused by noise. They require the specification of models. In the following section, the model concept from mathematical logic is complemented with a model concept that defines models as tools for representing and understanding the world.

2.4 Models

Models are central elements of an understanding of science. Giere (1999) concludes that "models play a much larger role in science than even the most ardent enthusiasts for models have typically claimed." Perhaps the most influential paper that describes the meaning and use of models in mathematics (especially in mathematical logic) and empirical sciences is Suppes (1969a). Based on Tarski's definition: "A possible realization in which all valid sentences of a theory T are satisfied is called a model of T" (Tarski 1953), Suppes asserts "that the meaning of the concept of model is the same in mathematics and the empirical science," although the concept of model is used in a different manner in these disciplines (Suppes 1969b). The concept of model used by mathematical logicians is the basic and fundamental concept of model needed for an exact statement of any branch of empirical science.

Logicians examine models that consist of abstract entities, e.g., geometrical objects. Suppes's proposition that there is no difference in the concept of a model in empirical science and in mathematics is based on the consideration that these objects could be physical objects. Or as David Hilbert stated decades ago: "Man muss jederzeit anstelle von Punkten, Geraden und Ebenen Tische, Stühle und Bierseidel sagen können."[1] (*It has to be possible to say tables, chairs, and beer mugs instead of points, lines, and planes at any time.*) Suppes establishes a relationship between theories (sets of axioms) and models (sets of objects satisfying the axioms). A model provides an interpretation of a set of uninterpretated axioms, called interpretative models.

By introducing error terms, model descriptions such as mathematical formulas can be interpreted as hypotheses about real-world systems. Hypotheses can be tested based on evidence obtained by examining real-world objects, i.e., by performing experiments. Higher-level models are not compared directly with data, but with models of data that rank lower in the hierarchy of models. In-between must be a model of experiments (Suppes 1969b). Giere summarizes Suppes' hierarchy of models as follows:

[1] This famous quotation cannot be found in Hilbert's publications. Walter Felscher wrote: "I have looked through Hilbert's articles on geometry, as well as through those on the foundations of mathematics, but nowhere did I find formulations mentioning Tische, Stuehle, Bierseidel. So the dictum seems indeed to be only such, not a quotation documentable in Hilbert's own publications" (Felscher 1998).

1. theoretical principles
2. theoretical models
3. models of experiments
4. models of data
5. data

Theoretical models describe how a substantive inquiry can be divided into local questions that can be probed. Experimental models are used to relate questions to canonical questions about the particular type of experiment and how to relate data to these experimental questions. Models of data describe how raw data can be generated and modeled so as to put them into a canonical form. In addition, they describe how to check if the data generation satisfies assumptions of the experimental models.

Following Suppes and Hilbert, models consist of abstract entities that could be in principle physical objects. But is this view of models adequate for physical objects? Giere (1999) discusses maps, diagrams, and scale models (models of the solar system or model houses) as representational models. He characterizes Suppes's models as *instantial* in contrast to his understanding, which is *representational*. Representational means that models are tools for representing the world for specific purposes, and not primarily providing means for interpreting formal systems. The representational view is related to the systems analysis process that requires a discussion of the context in which the need for a model arises before the subject of models and modeling is introduced (Schmidt 1986).

From the instantial view of models there is a direct relationship between linguistic expressions and objects. A circle can be defined as the set of points that have a constant distance (radius) from one specific point. The mathematical object "circle" can be linked to the linguistic expression without loss. But physical objects that cannot be observed precisely, and cannot be defined as exactly as theoretical objects, require a different conception.

Testing the fit of a model with the world, the model is compared with another model of data. It is not compared to data. And, scientific reasoning is "models almost all the way up and models almost all the way down" (Giere 1999). A significant use of models appears in mathematical statistics, where models are used to analyze the relation between theory and experimental data (Suppes 1969a). We will concentrate on the usage of models in mathematical statistics. The following section presents models of statistical testing in the framework of the new experimentalism.

2.5 The New Experimentalism

A naive description of the Popperian paradigm how scientific theories are constructed is based on three assumptions (Chalmers 1999):

1. Generalizations are based on a huge number of observations.

2. Observations have been repeated under a huge number of varying conditions.
3. No statement violates commonly accepted principles.

The vagueness of the term "huge number" is not the only problem of this approach. Sometimes, only a small number of observations is necessary to understand an effect: The destructive power of the atomic bomb has fortunately been demonstrated only rarely in the last decades.

The second assumption requires additional knowledge to differentiate between significant and insignificant variations (the colors of the experimenter's socks should have no significant impact on the experimental outcome—*ceteris paribus conditions* of an experiment are usually not incorporated in models). This additional knowledge can be concluded from well-known facts. As these well-known facts have to be justified and depend on the specification of further additional knowledge, this leads to an infinite regress.

Popper demands that theories should be falsifiable. "Good" theories survive many attempts of falsification. However, it remains unclear whether it is the theory or the additional assumptions, which are necessary to construct the theory, that are responsible for its falsification. Also Popper does not provide positive characterizations that would allow the discovery of survivable theories.

This is where the new experimentalism comes into play: The new experimentalists are seeking a scientific method, not through observation (passive), but through experimentation (active). They claim that experiments have negative and positive functionality. The experimenter can learn from mistakes because he has some knowledge of their causes.

Experiments live a life of their own; they do not necessarily require complex theories and are theory-neutral. Faraday's electric motor is one famous example to illustrate how experiments can be performed successfully and independently from high-level theories: "Faraday had no theory of what he had found" (Hacking 1983, p. 211).

The new experimentalists are looking for scientific conclusions that can be validated independently from complex abstract theories. Experiments can verify and falsify assertions and identify formerly unknown effects. Experimental results are treated as samples from the set of all possible results that can be drawn from experiments of this type. Error statistics are used to assign probabilities to sampling events. A theory is supported if predictions based on this theory have been proven.

Science is seen as the growth of experimental knowledge. The new experimentalists provide substantial means that enable experimenters to derive experimental knowledge independently from theory. One example how learning from experiments can be carried out will be detailed next.

2.5.1 Mayo's Models of Statistical Testing

Mayo (1996) attempts to capture the implications of the use of models in mathematical statistics in a rigorous way. A statistically modeled inquiry consists of three components (Mayo 1983), see Fig. 2.2:

(1) A *scientific claim* $\mathcal{C}$ can be modeled in terms of *statistical hypotheses* about a population. *Statistical models of hypotheses* enable the translation of scientific claims into claims about a certain population of objects. A *random variable* Y with *probability distribution* P describes a quantity of interest. *Statistical hypotheses* are hypotheses about the value of parameters of the probability distribution P, e.g., the mean μ of Y. A *probability model*

$$M(\mu) = \{\Pr(Y|\mu), \mu \in \Omega\} \tag{2.1}$$

describes Y, where Ω denotes the *parameter space.*

(2) Experimental *testing rules* can be used to model the observational analysis of the statistical hypotheses. The *sample space* $\mathcal{Y}$ is the set of possible experimental results $y = (y_1, \ldots, y_n)$. Each y_i is the realization of an independent random variable Y_i that is distributed according to $M(\mu)$. The probability of an experimental result $Pr(y|\mu)$ can be determined. Based on a *test statistic* T, a *testing rule* RU maps outcomes in $\mathcal{Y}$ to various claims about the model of hypotheses $M(\mu)$:

$$RU : \mathcal{Y} \rightarrow M(\mu). \tag{2.2}$$

A *statistical model of experimental tests* $ET(\mathcal{Y})$ is the triple

$$(\mathcal{Y}, \Pr(x|\mu), RU).$$

(3) The actual *observation* $\mathcal{O}$ can be modeled in terms of a statistical sample from the population. A statistical model of data models an empirical observation $\mathcal{O}$ as a particular element of $\mathcal{Y}$. It includes various sample properties of interest, for example, the mean.

As we will see in the following, it is crucial to leave some room between statistical and scientifical conclusions. A statistically significant conclusion is not automatically scientifically meaningful (Cohen 1995).

2.5.2 Neyman–Pearson Philosophy

The classical Neyman–Pearson theory of testing requires the determination of the region of the parameter space Ω in the hypothesis model $M(\mu)$ that will be associated with the null hypothesis H and the determination of the region that will be associated with an alternative hypothesis J. Applications of the NPT rules lead to a rejection of H and an acceptance of J, or vice versa.

Before the sample is observed, the experimental testing model $ET(\mathcal{Y})$ specifies which of the outcomes in $\mathcal{Y}$ should be taken to reject H. These values

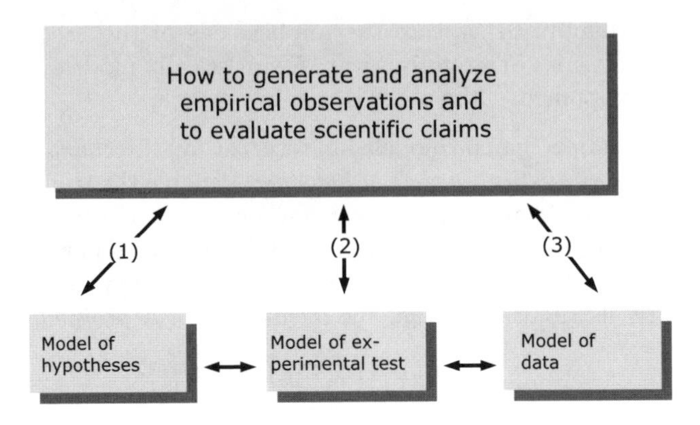

Fig. 2.2. Models of statistical testing. Mayo (1983) develops a framework that permits a delinearization of the complex steps from raw data to scientific hypotheses. Primary questions arise when a substantive scientific question is broken down into several local hypotheses. Experimental models link primary questions based on the model of hypotheses to questions about the actual experiment. Data models describe how raw data are transformed before. Not the raw data, but these modeled data are passed to the experimental models. Mayo (1983) describes three major *metastatistical tasks*: "(1) relating the statistical hypotheses [...] and the results of testing them to scientific claims; (2) specifying the components of experimental test [...]; and, (3) ascertaining whether the assumptions of a model for the data of the experimental test are met by the empirical observations [...]"

form the *critical region* (CR). A *type-I error* occurs if a true H is rejected, a *type-II error*, when a false H is accepted, and α and β denote the corresponding error probabilities. These error probabilities can be controlled by specifying a test statistic T and a testing rule RU that defines which of the possible values of T are mapped to the critical region. In general, a (test) statistic is a function of the observed random variables obtained in a random sample. It can be used as a point estimate for a population parameter, for example, the mean, or as a test statistic in hypothesis testing. A testing rule RU with $Pr(T \in \mathrm{CR}|\mu \in \Omega_H) \leq \alpha$ and $1 - \Pr(T \in \mathrm{CR}|\mu \in \Omega_J) \leq \beta$ can be specified, because the probability that $T \in \mathrm{CR}$ can be determined under various values of μ. The event $\{RU \text{ rejects } H\}$ can be identified with $\{T \in \mathrm{CR}\}$. It follows

$$Pr(RU \text{ rejects } H|H \text{ is true }\} = \Pr(T \in \ \mathrm{CR}\ |\mu \in \Omega_H\} \leq \alpha. \tag{2.3}$$

$$Pr(RU \text{ accepts } H|J \text{ is true }\} = 1 - \Pr(T \in \ \mathrm{CR}\ |\mu \in \Omega_J\} \leq \beta. \tag{2.4}$$

The simultaneous minimization of α and β is a conflicting goal. The two types of error are inversely related to each other, and it is impossible to minimize both of them simultaneously without increasing the sample size. Usually, the *significance level* α of a test is selected first. The significance level of a test can also be referred to as the *size*. In a second step a test with a small β value is chosen. The best NP test is the one that minimizes the probability of type-II errors for all possible values of μ under the alternative J. NP tests are objective, because they control the error probabilities independently from the true value of μ.

It is fundamental to be aware of two different interpretations of statistical tests. In a similar manner as Cox and Hinkley distinguish "the theory of statistical methods for the interpretation of scientific and technological data" and statistical decision theory, "where statistical data are used for more or less mechanical decision making"(Cox & Hinkley 1974), Mayo describes *statistical tests* (ST) as rules of inductive behavior and learning tools.

(ST-2.1) *Statistical tests as rules of inductive behavior.* Statistical tests can be interpreted as *rules of inductive behavior* and provide an automatic rule for testing hypotheses (Mayo 1996, p. 368). "To accept a hypothesis H means only to decide to take action A rather than action B"(Neyman 1950, p. 258). These behavioristic models of tests and the related automatisms are adequate means "when circumstances force us to make a choice between two courses of action"(Neyman 1950, p. 258). Automatic test rules play an important role for selection procedures of search algorithms under uncertainty, for example, the threshold selection scheme for direct search algorithms introduced in Chap. 8.

(ST-2.2) *Statistical tests as learning tools.* Mayo (1983) reformulates the Neyman–Pearson theory of testing and argues that it provides an objective theory of statistics. The control of error probabilities provides means to evaluate what has been learned from the results of a statistical test. Mayo describes tests as *learning tools*: "A test teaches about a specific aspect of the process that produces the data" (Mayo 1996, p.382).

It might be useful to present an example that uses simplified assumptions, e.g., common known variances, to explain the objective theory of statistical testing.

Example 2.2 (PSO swarm-size). In Example 2.1 a random sample was drawn from each of the two populations to determine whether or not the *difference between the two population means* is equal to δ. The two samples are independent, and each population is assumed to be normally distributed with common known standard deviation σ. The question is whether the increased swarm size improves the performance of the PSO. This can be formulated as the scientific claim C_1:

PSO(40) has a better (smaller) mean best fitness value than PSO(20).

(A) *Statistical hypothesis.* The model $M(\delta)$ is in the class of normal distribution, that is, it is supposed that the standard deviation σ is known

($\sigma = 160$) and that the variability of $\overline{Y}_i$, the mean best fitness value from n experiments, can be modeled by a normal distribution, $i = 1, 2$. If PSO(40) does not improve the performance, the difference δ between the two population means μ_1 and μ_2 would be zero. On the other hand, if $\mathcal{C}_1$ is true, δ will be greater than zero. The hypotheses read:

$$\begin{aligned} &\text{Null hypothesis } H: \delta = 0 \text{ in } \mathcal{N}(\delta, \sigma^2). \\ &\text{Alternative hypothesis } J: \delta > 0 \text{ in } \mathcal{N}(\delta, \sigma^2). \end{aligned} \tag{2.5}$$

(B) *Specifying the components of an experimental testing model (ET_1).* The vector $y_i = (y_{i1}, \ldots, y_{in})$ represents n observations from the ith configuration, and $\overline{y}_i$ denotes the ith sample mean, $i = 1, 2$. The experimental test statistic is $T = \overline{Y}_{12} = \overline{Y}_1 - \overline{Y}_2$, and its distribution under H is $\mathcal{N}(0, 2\sigma^2/n)$. The *upper α percentage point of the normal distribution* is denoted as z_α, for example, $z_{0.05} = 1.64$, or $z_{0.01} = 2.33$. As the number of observations was set to $n = 50$, it follows that the value of the *standard error* is $\sigma_{\overline{d}} = \sigma_{\overline{y}_1 - \overline{y}_2} = 160\sqrt{2/50} = 32$. The significance level of the test was $\alpha = 0.01$, thus $z_\alpha = z_{0.01} = 2.33$. So the test rule *RU* is

$$T: \text{ Reject } H: \delta = 0 \text{ if } T = \overline{Y}_1 - \overline{Y}_2 \geq 0 + z_\alpha \cdot \sigma_{\overline{d}} = 74.44.$$

(C) *Sample data.* The average performance $\overline{y}_1$ of $n = 50$ runs of PSO(20) is 108.02, whereas the average performance $\overline{y}_2$ of $n = 50$ runs of PSO(40) is 56.29. The difference $\overline{d} = \overline{y}_1 - \overline{y}_2$ is 51.73. Since this value does not exceed 74.44, *RU* does not reject H. ■

Example 2.2 shows a typical application of the Neyman–Pearson theory of testing. NPT has been under attack for many years. We will discuss important objections against NPT in the following and present an approach (NPT*) developed by Mayo to avoid these difficulties.

2.5.3 The Objectivity of NPT: Problems and Misunderstandings

The specification of a significance level α or error of the first kind is a crucial issue in statistical hypothesis testing. Hence, it is not surprising that attacks on the objectivity of NPT start with denying the objectivity of the specification of α. Why do many textbooks recommend a value of 5%?

Problem 2.1. "In no case can the appropriate significance level be determined in an objective manner" (Rubin 1971).

Significance tests are tests of a null hypothesis. The significance level is often called the *p*-value. Another question that attacks the Neyman–Pearson theory of statistical testing refers to the *p*-value.

Problem 2.2. In the context of a statistical test, does a given *p*-value convey stronger evidence about the null hypothesis in a larger trial than in a smaller trial, or vice versa? (Gregoire 2001).

Prima facie, one would answer that the p-value in a larger trial conveys stronger evidence. But the chance of detecting a difference increases as the sample size grows.

Although whole books devoted to these questions have been written, for example, *The Significance Tests Controversy* (Morrison & Henkel 1970), this controversy has not been recognized outside the statistical community. Gigerenzer states that there are no textbooks (written by and addressed to nonstatisticians like psychologists) that explain differences in opinion about the logic of inference: "Instead, readers are presented with an intellectually incoherent mix of Fisherian and Neyman–Pearsonian ideas, but a mix presented as a seamless, uncontroversial whole: the logic of scientific inference" (Gigerenzer 2003). The following section discusses the definition of the p-value to clarify these questions.

2.5.4 The Objectivity of NPT: Defense and Understanding

Significance and the p-Value

Sometimes scientists claim that their results are significant because of the small p-value. The p-value is taken as an indicator that the null hypothesis is true (or false). This is as wrong as claiming that the movement of the leaves in the trees causes windstorms in autumn. The p-value is

$$p = \Pr\{ \text{ result from test-statistic, or greater} \mid \text{null model is true } \},$$

and not a measure of

$$p = \Pr\{ \text{ null model is true} \mid \text{test-statistic } \}.$$

Therefore, the p-value has "no information to impart about the verity of the null model itself" (Gregoire 2001). The p-value is not related to any probability whether the null hypothesis is true or false. J. Neyman and E.S. Pearson[2] proposed a framework of acceptance and rejection of a statistical hypothesis instead of a framework of significance. A significant result is a beginning, not an end. "Eating beans and peas 'significantly' decreases the probability of getting lung cancer. But why on Earth?" (Hacking 2001). The specification of a p-value depends on the context in which the need for an experimental test arises. Researchers can judge the possible consequences of a wrong decision. Mayo comments on problem 2.1:

Answer (to Problem 2.1). The fact that the same data lead to different conclusions depending on the specification of α is entirely appropriate when such specifications are intended to reflect the researcher's assessment of the consequences of erroneous conclusions (Mayo 1983, p. 315).

---

[2] Egon Pearson should not be confused with his father Karl, who proposed a different philosophy.

The specifications, which are made in using NPT, are what allows tests to avoid being sterile formal exercises.

The misconception is that NPT functions which test scientific claims directly, cf. arrow (1) in Fig. 2.2, and functions from which the statistical hypotheses are derived, arrow (2), are collapsed. This misconception can be avoided if some room is left between the statistical and the scientific conclusion.

How can statistical tests in a scientific inquiry provide "means of learning"? Being able to use the distribution of the test statistic T, error probabilities can be objectively controlled. This provides objective scientific knowledge. Detecting discrepancies between the correct and the hypothesized models enables learning about phenomena, for example, "Is the actual value of δ positively discrepant from the hypothesized value δ_0?"

Severity

From the distribution of the experimental test statistic T it is known that it is possible for an observed $\overline{d}$ to differ from a hypothesis δ_0, when no discrepancy exists (between δ, the true difference in means, and δ_0), or vice versa. A suitable choice of T enables the variation of the size of the observed differences, in a known manner, with the size of the underlying discrepancies. A test can be specified so that it will not very often classify an observed difference as significant (and hence reject H) when no discrepancy of scientific importance is detected, and not very often fail to do so (and so accept H) when δ is importantly discrepant from δ_0. This principle can be expressed in terms of the *severity requirement* (SR). It leads to NPT*, Mayos's extension of NPT.

(SR) An experimental result e constitutes evidence in support of a hypothesis H just to the extent that:
(SR-1) e fits H, and
(SR-2) H passes a severe test with e.

The requirement (SR-2) can be further specified by means of the *severity criterion*: A hypothesis H passes a severe test T with outcome e just in case the probability of H passing T with an outcome such as e (i.e., one that fits H as well as e does), given H is false, is very low. Staley (2002) additionally formulates a *severity question* (SQ) as follows: The severity question needs to be addressed to determine whether a given experimental outcome e is evidence for a given hypothesis H:

(SQ) How often would a result like this occur, assuming that the hypothesis is false?

Note that the severity requirements are not related to the error of the second kind (β error). Based on the severity requirements, we attempt to show how NPT* can be used to objectively interpret statistical conclusions from experimental tests.

An Objective Interpretation of Rejecting a Hypothesis

The question "Are the differences real or due to the experimental error?" is central for the following considerations. The metastatistical evaluation of the test results tries to determine whether the scientific import is misconstrued. A positive difference between the true and the observed value of less than one or two standard deviation units is quite often caused by experimental error. Thus, small differences may often erroneously be confused with effects due to real differences. This problem can be tackled by selecting a small α error, because only observed differences as large as $z_\alpha \sigma_{\overline{d}}$ are taken to reject H (Fig. 2.3). But choosing a small α value alone is not sufficient, because the

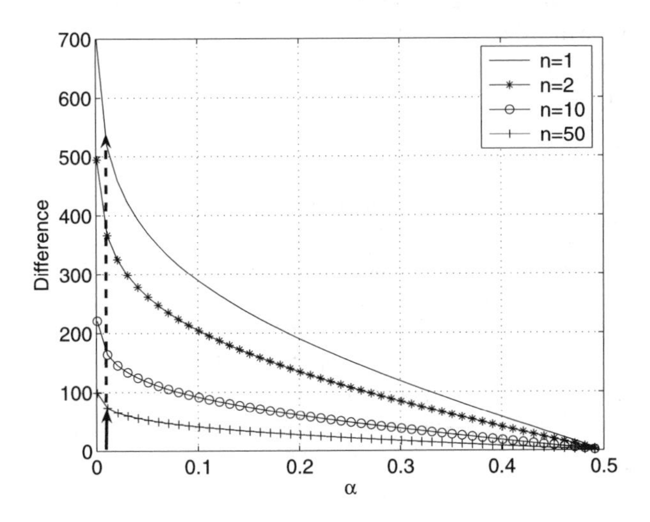

Fig. 2.3. Influence of the sample size n on the test result. Plot of the difference $z_\alpha \cdot \sigma_{\overline{d}}$ versus α, with $\sigma = 160$. The *arrows* point out the influence of the sample size for a given error of the first kind ($\alpha = 0.01$) on the difference: If $n = 1$ the difference $\overline{d} = \overline{Y}_1 - \overline{Y}_2$ must exceed 526.39 to reject the null hypothesis. If more experiments are performed ($n = 50$), this difference must exceed only the seventh part of this value: 74.44. To demonstrate that there is no difference in means, experimenters can reduce the sample size n. Otherwise, increasing the sample size n may lead to a rejection of the null hypothesis

standard error $\sigma_{\overline{d}}$ depends on the sample size n. A *misconstrual* (MI) is a wrong interpretation resulting from putting a wrong construction on words or actions. Mayo (1983) describes the first misconstrual as follows:

(MI-2.1) A test can be specified that will almost give rise to an average difference $\overline{d} = \overline{y_1} - \overline{y_2}$ that exceeds δ_0 by the required difference $z_\alpha \sigma_{\overline{d}}$, even if the underlying δ exceeds δ_0 by as little as one likes.

This can be accomplished by selecting an appropriately large sample size n. If one is allowed to go on sampling long enough ($n \to \infty$), then even if the null hypothesis H is true, one is assured of achieving a statistically significant difference from H. The likelihood that we can detect a difference (power) in the test increases.

Example 2.3 (Sample size). Consider $Y_1 \sim \mathcal{N}(100, 5)$ and $Y_2 \sim \mathcal{N}(110, 5)$. Choosing a test with $n = 50$ will almost give rise to a $\overline{Y}_{21}$ that exceeds δ_0 by the required difference. Note that we wish to reject H if one mean is larger than the other. Therefore we are interested in the difference $\overline{Y}_{21} = \overline{Y}_2 - \overline{Y}_1$. Next, consider $Y_1 \sim \mathcal{N}(100, 5)$ and $Y_3 \sim \mathcal{N}(100.1, 5)$. If the sample size is increased, say $n = 5000$, a similar result can be obtained. ■

In the following (Figs. 2.6–2.8), graphical tools to make these dependencies understandable will be developed.

Summarizing, the product $z_\alpha \sigma_{\overline{d}}$ can be modified by changing the sample size n or the error α:

$$\lim_{\alpha \to 0} z_\alpha = \infty. \tag{2.6}$$

$$\lim_{n \to \infty} \sigma_{\overline{d}} = 0. \tag{2.7}$$

NPT allows misconstruals of the scientific import if a rejection of H is automatically taken to indicate that the scientific claim $\mathcal{C}$ is true. Even scientifically unimportant differences are classified as important because the test is too sensitive or powerful. When A and B are different treatments with associated means μ_A and μ_B, μ_A and μ_B are certain to differ in some decimal place so that $\mu_A - \mu_B = 0$ is known in advance to be false (Cohen 1990; Tukey 1991).

The Observed Significance Level

The frequency relation between a rejection of the null hypothesis H and values of the difference in means, δ, is important for the interpretation of the rejection. To interpret the rejection of H, Mayo introduces the *observed significance level*

$$\alpha_{\overline{d}}(\delta) = \alpha(\overline{d}, \delta) = \Pr(\overline{Y_1} - \overline{Y_2} \geq \overline{d} | \delta). \tag{2.8}$$

Hence, $\alpha_{\overline{d}}(\delta)$ is the area under the normal curve to the right of the observed $\overline{d}$, as illustrated in Fig. 2.4. If we set $\delta_0 = 0$, then $\alpha_{\overline{d}}(\delta_0)$ is the frequency of an error of the first kind. If $\alpha_{\overline{d}}(\delta_0) \leq$ "the preset significance level of the test *RU*," then *RU* rejects H with $\overline{d}$. Rejecting H with *RU* is a good indicator that $\delta > \delta_0$ to the extent that such a rejection is not typical when δ is as small as δ_0. If any and all positive discrepancies from δ_0 are regarded as scientifically important, then a small $\alpha_{\overline{d}}(\delta)$ value ensures that construing such a rejection as indicating a scientifically important δ does not occur very often. Small $\alpha_{\overline{d}}(\delta)$ values do not prevent an *RU* rejection of H from often being misconstrued

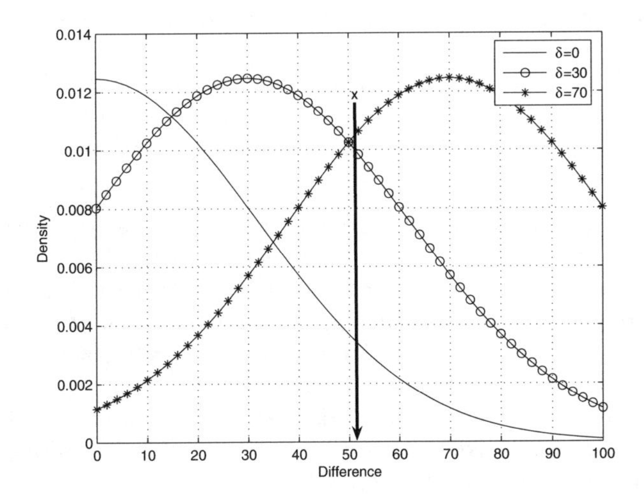

Fig. 2.4. Observed difference and three hypothetical differences. Difference in means for $n = 50$ samples and standard deviation $\sigma = 160$. The value from the test statistic $\overline{d} = 51.73$ remains fixed for varying means δ_i of different distributions associated with the null hypotheses H_i, $i = 1, 2, 3$. The figure depicts the probability density functions of the associated normal distributions for three different means: $\delta_1 = 0$, $\delta_2 = 30$, and $\delta_3 = 70$. To interpret the results, consider a hypothetical difference in means of $\delta_2 = 30$: The observed significance level $\alpha_{\overline{d}}(\delta_2)$ is the area under the normal curve to the right of $\overline{d}$. The value $\alpha_{51.75}(30) = 0.25$ is quite large and therefore not a good indication that the true difference in means is as large as $\delta_2 = 30$

when relating it to the scientific claim $\mathcal{C}$, if some δ values in excess of δ_0 are still not considered scientifically important.

Regard the values of $\alpha_{\overline{d}}(\delta')$ for $\delta' \in \Omega_J$. An *RU* rejection with $\overline{d}$ successfully indicates that $\delta > \delta'$ if $\alpha_{\overline{d}}(\delta')$ is small. If $\alpha_{\overline{d}}(\delta')$ is fairly large, then such a rejection is the sort of event that fairly frequently occurs when $\delta \leq \delta'$.

To relate the statistical result to the scientific import, Mayo proposes to define δ_{un}:

$$\delta_{\text{un}} = \text{ the largest scientifically unimportant value in excess of } \delta_0. \tag{2.9}$$

If $\alpha_{\overline{d}}(\delta_{\text{un}})$ is large, then the statistical result is not a good indication that the scientific claim is true. In addition to δ_{un}, we can define δ^{α}, the *inversion of the observed significance level* function as:

$$\delta^{\alpha} = \text{ the value of } \delta \text{ in } \Omega \text{ for which } \alpha_{\overline{d}}(\delta) = \alpha. \tag{2.10}$$

Example 2.4. Consider a sample size of $n = 50$. If $\delta_{\text{un}} = 30$, then rejecting H with *RU* cannot be taken as an indication that the scientific claim "PSO(40) outperforms PSO(20)" is true. The arrow in Fig. 2.5 illustrates this situation. The observed significance level $\alpha_{\overline{d}}(30) = 0.25$ is not a strong indication that δ

exceeds 30. However, if the sample size is increased ($n = 500$), then $\alpha_{\overline{d}}(30) = 0.05$ is small.

Consider Example 2.1 and an observed significance level $\alpha = 0.5$. Then the value of the inversion of the observed significance level function is $\delta^{0.5} = 51.73$. As $\delta^{0.027} = \overline{y} - 2\sigma_{\overline{d}} = 31.49$ ($n = 500$), hence an *RU* rejection is an indication of $\delta > \delta^{0.027}$. ■

But are these results good indications that one is observing a difference $\delta > 0$ that is also scientifically important? This problem is outside the domain of statistics. Its answer requires the specification of a scientifically important difference, a reasonable sample size, and an acceptable error of the first kind, cf. statement 2.5.4. The $\alpha_{\overline{d}}(\delta)$ function provides a nonsubjective tool for understanding the δ values, a metastatistical rule that enables learning on the basis of a given *RU* rejection. As the examples demonstrate, NPT* tools enable the experimenter to control error probabilities in an objective manner. The situation considered so far is depicted in Fig. 2.5.

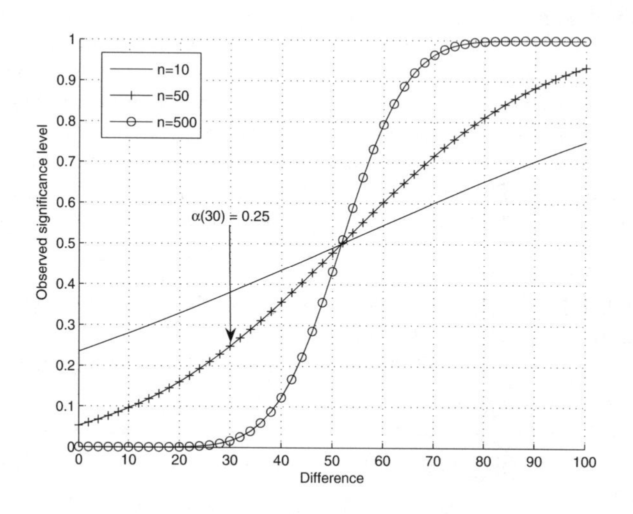

Fig. 2.5. Plot of the observed significance level $\alpha_{\overline{d}}(\delta)$ as a function of δ, the possible true difference in means. Lower $\alpha_{\overline{d}}(\delta)$ values support the assumption that there is a difference as large as δ. The measured difference is $\overline{d} = 51.73$, the standard deviation is $\sigma = 160$, cf. Example 2.1. The *arrow* points to the associated value of area under the normal curve to the right of the observed difference $\overline{d}$, as shown in Fig. 2.4. Each point of the three curves shown here represents one single curve from Fig. 2.4. The values can be interpreted as follows: Regard $n = 50$. If the true difference is (a) 0, (b) 51.73, or (c) 100, then (a) $H : \delta = 0$, (b) $H : \delta = 51.73$, or (c) $H : \delta = 100$ is wrongly rejected (a) 5%, (b) 50%, or (c) 95% of the time

An Objective Interpretation of Accepting a Hypothesis

In a similar manner as rejecting H with a test that is too sensitive may indicate scientifically important δ's have been found, accepting a hypothesis with a test that is too insensitive may fail to indicate that no important δ's have been found. This can be defined as the second misconstrual:

(MI-2.2) A test can be specified that will almost give rise to an average difference $\overline{d}$ that does not exceed δ_0 by the required difference $z_\alpha \sigma_{\overline{d}}$, even if the underlying δ exceeds δ_0 by as much as one likes.

To interpret the acceptance of H with RU, Mayo defines

$$\beta_{\overline{d}}(\delta) = \beta(\overline{d}, \delta) = \Pr(\overline{Y_1} - \overline{Y_2} \leq \overline{d} | \delta), \tag{2.11}$$

and δ^β as the value of δ in the parameter space Ω for which $\beta_{\overline{d}}(\delta) = \beta$, and finally

$$\delta_{\text{im}} = \text{the smallest scientifically important } \delta \text{ in excess of } \delta_0. \tag{2.12}$$

Learning

NPT* accomplishes the task of objectively interpreting statistical results. The testing rule RU requires assumptions on the distribution of the underlying empirical observations $\mathcal{O}$. This is seen as part of task (3), depicted as arrow (3) in Fig. 2.2. For example, one has to verify that the sample observation $\mathcal{O}$ can be modeled as the result of n independent observations of a random variable distributed according to the probability model $M(\delta)$. The assumption of independence can be checked using various goodness-of-fit tests. The learning function of tests may be accomplished even if the test assumptions are not satisfied precisely. Mayo claims that NPT methods are robust, and NPT* makes this robustness explicit.

To present a comprehensive example, we assumed a population that follows a Gaussian distribution with known variance. Based on the bootstrap, which will be detailed in Sect. 3.2, we are able to use our approach independently from any assumptions on the underlying distribution.

Example 2.5 (Rejecting a hypothesis). Consider the situation depicted in Fig. 2.6. The experimental test statistic $T = \overline{Y_1} - \overline{Y_2}$ is based on samples $Y_1 \sim \mathcal{N}(110, 5)$ and $Y_2 \sim \mathcal{N}(100, 5)$. The plot of the observed significance (Fig. 2.6 on the left) indicates that one is observing a difference $\delta > 0$, and that this difference is not due to an increased sample size n alone. The values from Table 2.1 reflect this assumption: The observed significance levels for 1 and 2 standard error units, $\alpha_{\overline{d}}(\sigma_{\overline{d}})$ and $\alpha_{\overline{d}}(2\sigma_{\overline{d}})$, are small. This case will be referred to as RE-2.1 in the remainder of this book. ■

Example 2.6 (Accepting a hypothesis). The curves of the observed significance level $\alpha_{\overline{d}}$ change their shape significantly if the true difference in means is very small, i.e., if $Y_1 \sim \mathcal{N}(100.1, 5)$ and $Y_2 \sim \mathcal{N}(100, 5)$. Figure 2.7 depicts this situation. Only a large sample size, i.e., $n = 5000$, is able to detect this difference; smaller sample sizes, i.e., $n = 10$, or $n = 50$, do not indicate a difference in means. Note that in addition to the absolute $\alpha_{\overline{d}}$ values, the slope (the rate of change) is of importance. Figure 2.7 gives no reason to reject the null hypothesis. This case will be referred to as RE-2.2 in the remainder of this work.

The corresponding plot that gives reason to accept the null hypothesis is shown in Fig. 2.8. Consider a sample size of $n = 5000$: The corresponding curve shows that we can safely state "there is no difference in means as large as $\delta = 0.4$." Figure 2.8 (right) shows a similar situation with reduced noise levels ($Y_1 \sim \mathcal{N}(100.1, 1)$ and $Y_2 \sim \mathcal{N}(100, 1)$). Regarding $n = 5000$, we can safely state that "there is no difference in means as large as $\delta = 0.2$." This case will be referred to as AC-2.1 in the remainder of this work. ∎

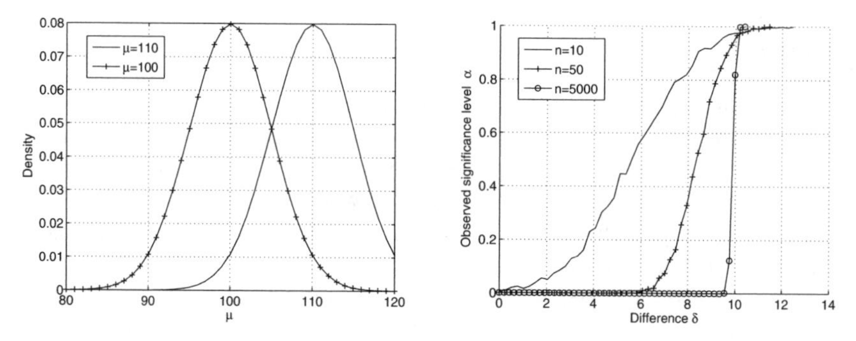

Fig. 2.6. Case RE-2.1. Rejecting a hypothesis. Density and observed significance level plots. $Y_1 \sim \mathcal{N}(110, 5)$, $Y_2 \sim \mathcal{N}(100, 5)$. This difference is meaningful and should be detected. These cases correspond to the configurations R1–R3 from Table 2.1

Answer (to Problem 2.2). Plots of the observed significance level are non-subjective tools for understanding the functional relationship between sample size and p-value.

After discussing the objective interpretation of accepting or rejecting a hypothesis, it is important to note that experiments consists of several tests. We have described one basic procedure only. In general, the learning process requires numerous statistical tests, and the problem is broken up into smaller pieces. "One is led to break things down if one wants to learn" (Mayo 1997, p. 254). In contrast to the no peeking rule (GL-2.3) from experimental algorithmics, Mayo allows peeking. Anyhow, experience shows that no experimenter can obey the no peeking rule without exception.

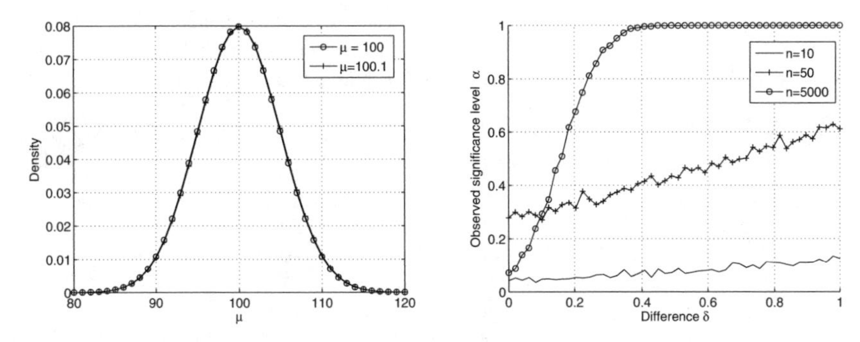

Fig. 2.7. Case RE-2.2. Rejecting a hypothesis. Density and observed significance level plots. $Y_1 \sim \mathcal{N}(100.1, 5)$, $Y_2 \sim \mathcal{N}(100, 5)$. This difference is not meaningful. These cases correspond to the configurations R4–R6 from Table 2.1

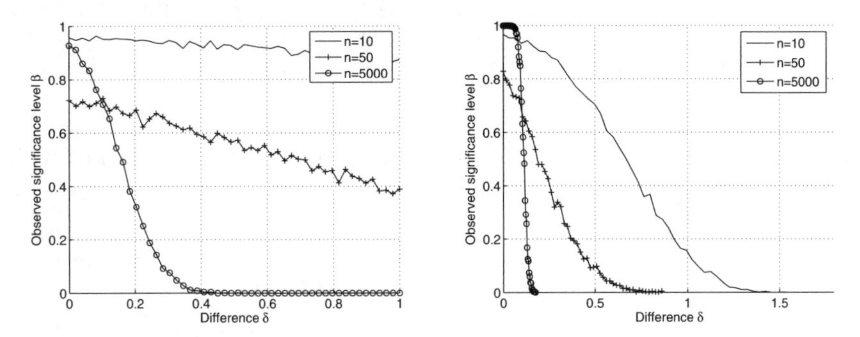

Fig. 2.8. Case AC-2.1. Accepting a hypothesis. Observed significance level plots. *Left*: $Y_1 \sim \mathcal{N}(100.1, 5)$, $Y_2 \sim \mathcal{N}(100, 5)$. The differences are probably not meaningful, because only large sample sizes produce small p-values. *Right*: $Y_1 \sim \mathcal{N}(100.1, 1)$, $Y_2 \sim \mathcal{N}(100, 1)$. These cases correspond to the configurations A1–A6 from Table 2.1

2.5.5 Related Approaches

Selvin (1970, p. 100) states that there is a difficulty in the interpretation of "significance" and "level of significance." The level of significance is the probability of wrongly rejecting the null hypothesis that there is no difference between two populations. The significance describes the scientific meaning of this difference.

In addition to this, the observed significance level is closely related to *consonance intervals* as introduced in Kempthorne & Folks (1971). Consonance intervals can be regarded as an inversion of significance tests: We ask for the degree of agreement of the parameters of a particular model with the data (Folks 1981; Kempthorne & Folks 1971). Given the data, the parameters of the model are evaluated. That is, observing 503 heads in 1000 coin

Table 2.1. Rejection and acceptance of hypotheses. The sample size n, the standard error $\sigma_{\overline{d}}$, the observed difference $\overline{d}$, the observed difference minus 1 and 2 $\sigma_{\overline{d}}$'s, and the values of the observed significance levels for 1 and 2 $\sigma_{\overline{d}}$'s are shown

Conf.	n	$\sigma_{\overline{d}}$	$\overline{d}$	$\overline{d}-\sigma_{\overline{d}}$	$\overline{d}-2\sigma_{\overline{d}}$	$\alpha_{\overline{d}}(\sigma_{\overline{d}})$	$\alpha_{\overline{d}}(2\sigma_{\overline{d}})$
$R1$	10	1.99	9.58	7.59	5.60	0	0
$R2$	50	0.94	10.66	9.72	8.79	0	0
$R3$	5000	0.1	9.89	9.79	9.69	0	0
$R4$	10	1.57	2.99	1.41	−0.16	–	–
$R5$	50	1.06	0.7	−0.36	−1.43	–	–
$R6$	5000	0.1	0.18	0.08	−0.02	0.29	0.67
Conf.	n	$\sigma_{\overline{d}}$	$\overline{d}$	$\overline{d}+\sigma_{\overline{d}}$	$\overline{d}+2\sigma_{\overline{d}}$	$\alpha_{\overline{d}}(\sigma_{\overline{d}})$	$\alpha_{\overline{d}}(2\sigma_{\overline{d}})$
$A1$	10	1.57	2.99	4.56	6.13	–	–
$A2$	50	1.06	0.7	1.76	2.83	–	–
$A3$	5000	0.1	0.18	0.28	0.38	0.71	0.33
$A4$	10	0.31	0.68	0.99	1.31	0.15	0.46
$A5$	50	0.21	0.22	0.43	0.65	0.48	0.14
$A6$	5000	0.02	0.12	0.14	0.16	1	1

tosses, the model "heads and tails are equally probable" is consonant with the observation.

2.6 Popper and the New Experimentalists

In a similar manner as Gigerenzer (2003) presents his tools to theories approach, Mayo suspects that Popper's falsification theory is well accepted by many scientists since it reflects the standard hypothesis testing principles of their daily practice. To clarify the difference between Mayo's NPT* approach and Popperian testing, the reader may consider the following quotation:

> Mere supporting instances are as a rule too cheap to be worth having: they can always be had for the asking; thus they cannot carry any weight; and any support capable of carrying weight can only rest upon ingenious tests, undertaken with the aim of refuting our hypothesis, if it can be refuted (Popper 1983).

Popper's focus lies on the rejection of hypotheses; his concept of severity does not include tools to support the acceptance of scientific hypotheses as introduced in Sect. 2.5. Popper states that the theoretician will try to detect any false theories; that is, he will "try to construct severe tests and crucial test situations" (Popper 1979). But Popper does not present objective interpretations (as Mayo does) for accepting and rejecting hypotheses.

Mayo explicitly states that it is important to distinguish Popperian severity from hers. Also Popper stresses the importance of severe tests: Hypothesis H passes a severe test with experimental result e if all alternatives to H

that have been tested entail not-e. But there are many not-yet-considered or not-yet-even-thought-of alternative hypotheses that also entail e. Why is this alternative objection not relevant for NPT*? Mayo comments:

> Because for H to pass a severe test in my sense it must have passed a severe test with high power at probing the ways H can err. And the test that alternative hypothesis H' failed need not be probative in the least so far as H's errors go. So long as two different hypotheses can err in different ways, different tests are needed to probe them severely (Mayo 1997, p. 251).

Hypothesis H has passed a severe test if the hypothesis itself has been tested. It is not sufficient—as claimed in the Popperian paradigm—that all alternative hypotheses to H failed and H was the only hypothesis that passed the test.

And What About Theory?

According to the new experimentalists, an experiment can have a life of its own. An experiment is independent of large-scale theory. This is an obvious contradiction to the Popperian view that theory precedes experiments and that there is no experiment without theory. While discussing the relationship between theory and experiment, Hacking comments on Popper's statement: "Theory dominates the experimental work from its initial planning to the finishing touches in the laboratory" (Popper 1959) with a counterexample that mentions Humphry Davy (1778–1829):

> Davy's noticing the bubble of air over the algae is one of these [obvious counterexamples]. It was not an "interpretation in the light of theory" for Davy had initially no theory. Nor was seeing the taper flare an interpretation. Perhaps if he went on to say, "Ah, then it is oxygen", he would have been making an interpretation. He did not do that (Hacking 1983).

We note additionally that a great many examples from the history of science can be mentioned as counterexamples to the Popperian view that theory dominates the experimental work. Davy experimented with gases by inhaling them and thus invented the laughing gas, nitrous oxide, without any theory. And, in opposition to existing theories, Davy showed that hydrochloric acid did not contain oxygen. One last example from Davy: Ice cubes melt when they are rubbed together—in contradiction to the caloric theory. Summarizing the discussion about theory from this chapter, we conclude: Theory can be described as wishful thinking. Moreover it is defined through consulting a dictionary as:

1. The analysis of a set of facts in their relation to one another.
2. An abstract thought: speculation (Merriam-Webster Online Dictionary 2004).

The latter definition is consonant with Hacking's suggestion not to differentiate only between theory and experiment, but to use a tripartite division instead: speculation, calculation, and experimentation (Hacking 1996). We will not enforce the gap between theory and practice any further, because we sympathize with the idea of this tripartite division that will be reconsidered in Chap. 9. The reader is also referred to Hacking's *Representing and Intervening* that details these ideas.

2.7 Summary

The results from this chapter can be summarized as follows:

1. The goal of this work is to lay the groundwork for experimental research in evolutionary computation.
2. Solely theoretical approaches to investigate, compare, and understand algorithms are not satisfactory from an experimenter's point of view.
3. Algorithm runs can be treated as experiments.
4. Guidelines from experimental algorithmics provide good starting points for experimental studies.
5. Experimental algorithmics is based on the Popperian paradigms:
   (a) There is no experiment without high-level theories.
   (b) Theories should be falsifiable.
6. We claim that:
   (a) There are experiments without high-level theories ("experiment can have a life of its own").
   (b) Popper's falsification should be complemented with validation.
7. The concept of the new experimentalism is transferred from philosophy to computer science, especially to evolutionary computation.
8. Models are central elements of an understanding of science.
9. Mayo introduced models of statistical testing that leave room between scientific claims and statistical hypotheses.
10. Hypothesis testing can be applied as an automatic rule (NPT) or as a learning tool (NPT*).
11. The approach presented here enables learning from experiments. Learning can be guided by plots of the observed significance against the difference, as shown in Fig. 2.5.

Example 2.2 was based on the assumption of known variances and normally distributed data. The following chapter introduces statistical tools that enable the application of NPT* methods even if the underlying distribution is unknown.

2.8 Further Reading

The modern theory of statistical testing presented in this chapter is based on Hacking (1983) and Mayo (1996). Morrison & Henkel (1970) discuss the significance test controversy.

3

Statistics for Computer Experiments

Like dreams, statistics are a form of wish fulfillment.
—Jean Baudrillard

This chapter discusses some methods from classical and modern statistics. The term "computational statistics" subsumes computationally intensive methods. They comprise methods ranging from exploratory data analysis to Monte Carlo methods. Data should be enabled to "tell their story". Many methods from computational statistics do not require any assumptions on the underlying distribution. Computer based simulations facilitate the development of statistical theories: 50 out of 61 articles in the theory and methods section of the Journal of the American Statistical Association in 2002 used Monte Carlo simulations (Gentle et al. 2004a).

The accuracy and precision of data may be limited due to noise. How can deterministic systems like computers model this randomness? Stochastic computer experiments, as performed in evolutionary computation, have to cope with a different concept of randomness than agricultural or industrial experiments. The latter face inherent randomness, whereas the former require methods to generate randomness. This is accomplished by generating sequences of *pseudorandom numbers*.

A sequence of infinite length is random if the quantity of information it contains is infinite too. If the sequence is finite, it is formally impossible to verify whether it is random or not. This results in the concept of pseudorandomness: Statistical features of the sequence in question are tested, i.e., the equiprobability of all numbers (Knuth 1981; Schneier 1996). Following this rather pragmatic approach, randomness and pseudorandomness will be treated equivalently throughout the rest of this work.

First, some basic definitions from hypothesis testing are introduced. Next, a bootstrap method to generate significance plots as shown in Fig. 2.5 is described. It provides an effective method to use the raw data without making any additional assumptions on the underlying distribution. The bootstrap can solve problems that would be too complicated for classical statistical techniques.

Then some useful tools for the analysis of computer experiments are presented. Common to all these methods is that they provide means to explain the variability in the data. Varying the exogenous strategy parameters of an optimization algorithm may cause a change in its performance. Therefore we will be able to find answers to the research goals RG 2.1–2.4.

Classical statistical methods such as the analysis of variance or regression analysis are common means to tackle these questions. These numerical techniques should be complemented with graphical tools such as histograms, box plots, or interaction plots, which belong to the standard repertoire of many statistical software packages. In particular, regression trees that are distribution-free methods to visualize structure in data have been proven useful in this analysis. This chapter concludes with an introduction of the basic definitions for DACE models, which can be seen as an extension of the classical approach from regression analysis.

A comprehensive introduction into statistical methods cannot be given here. Instead this chapter focuses on the basic ideas.

3.1 Hypothesis Testing

The following subsections introduce the basic definitions used for statistical hypothesis testing. Example 2.2 is reconsidered. Besides the three procedures presented here, many more test procedures exist. These tests can be classified according to known or unknown variances, equal or unequal sample sizes, and equal or unequal variances (Montgomery 2001). The z-test is presented first, because it was used in Example 2.2. In contrast to the z-test, where variances have to be known, in the t-test estimates of the variances are computed.

3.1.1 The Two-Sample z-Test

In Example 2.2 the performances of two algorithms, PSO(20) and PSO(40), respectively, were compared. The vector $y_i = (y_{i1}, \ldots, y_{in_i})$ represents the n_i observations from the ith algorithm, $\overline{y}_i$ denotes the ith sample mean, and σ_i^2 the associated variances. The distribution of the difference in means $\overline{Y}_{12} = \overline{Y}_1 - \overline{Y}_2$ is $\mathcal{N}(\overline{y}_1 - \overline{y}_2, \sigma^2(1/n_1 + 1/n_2))$, if the samples were drawn from independent normal distributions $\mathcal{N}_i(\mu_i, \sigma_i^2)$ with common variance $\sigma^2 = \sigma_i^2$, $i = 1, 2$. If σ^2 were known and

$$\mu_1 = \mu_2, \tag{3.1}$$

then

$$Z_0 = \frac{\overline{y}_1 - \overline{y}_2}{\sigma\sqrt{1/n_1 + 1/n_2}} \sim \mathcal{N}(0, 1).$$

Equation (3.1) is a statement or a statistical hypothesis about the parameters of a probability distribution, the null hypothesis: $H : \mu_1 = \mu_2$. The alternative hypothesis can be defined as the statement $J : \mu_1 \neq \mu_2$. The one-sided alternative hypothesis can be specified as $J : \mu_1 > \mu_2$. The procedure of formulating a hypothesis H, taking a random sample, computing a test statistic, and the acceptance of (or failure to accept) H is called *hypothesis testing*. The critical region contains the values that lead to a rejection of H. The *significance level* α is the probability of a type-I error for the test:

$$\alpha = \Pr(\text{type-I error}) = \Pr(\text{reject } H|H \text{ is true}). \tag{3.2}$$

The type-II error is defined as

$$\beta = \Pr(\text{type-II error}) = \Pr(\text{fail to reject } H|H \text{ is false}). \tag{3.3}$$

To determine whether to reject the null hypothesis $H : \mu_1 = \mu_2$, the value of the test statistic $T : \overline{d} = \overline{y}_1 - \overline{y}_2$ is compared to the normal distribution. If

$$\overline{d} \geq z_\alpha \sigma \sqrt{1/n_1 + 1/n_2},$$

where z_α is the *upper α percentage point* of the normal distribution, the null hypothesis would not be accepted in the classical *two-sample z-test*. When $\alpha = 0.01$, then z_α has the value 2.23. With $n_1 = n_2 = n$ and $\sigma = 160$ follows that $z_\alpha \sigma \sqrt{1/n_1 + 1/n_2} = 2.23 \cdot 160\sqrt{2/50} = 74.44$, as in Example 2.2.

The definition of the upper α percentage point of the normal distribution can be generalized to the case of more than one random variable. Let $(W_1, \ldots, W_s)$ have the s-variate normal distribution with mean vector zero, unit variances, and common correlation ρ. Then

$$\Pr\left(\max_{1 \leq i \leq s} W_i \leq Z_{s,\rho}^{(\alpha)}\right) = 1 - \alpha \tag{3.4}$$

defines the upper α equicoordinate critical point $Z_{s,\rho}^{(\alpha)}$ of this distribution (Bechhofer et al. 1995, p. 18).

3.1.2 The Two-Sample *t*-Test

If the variances of the populations are unknown, the sample variances

$$S_i^2 = \frac{\sum_{k=1}^{n_i} (y_{ik} - \overline{y}_i)^2}{n_i - 1}$$

can be used to estimate σ_i^2, $i = 1, 2$. The related test procedure is called the *two-sample t-test*. The upper α percentage point of the normal distribution is replaced by t_{α, n_1+n_2-2}, the *upper α percentage point of the t-distribution* with $n_1 + n_2 - 2$ degrees of freedom.

Let $S_p^2 = \left[(n_1 - 1)S_1^2 + (n_2 - 1)S_2^2\right] / (n_1 + n_2 - 2)$, the pooled variance, denote an estimate of the common variance σ^2. Then

$$t_0 = \frac{\overline{y}_1 - \overline{y}_2}{S_p^2 \sqrt{1/n_1 + 1/n_2}}. \tag{3.5}$$

If the null hypothesis H is true, t_0 is distributed as $t_{n_1+n_2-2}$, and $100(1-\alpha)$ percent of the values of t_0 lie in the interval $[-t_{\alpha/2,n_1+n_2-2}, t_{\alpha/2,n_1+n_2-2}]$, where $t_{\alpha,n}$ denotes the upper α percentage point of the t-distribution with n degrees of freedom.

The t-distribution with n_1+n_2-2 degrees of freedom is called the relevance distribution for the test statistic t_0. To reject H only if one mean is larger than the other ($\mu_1 > \mu_2$), the criterion

$$t_0 > t_{\alpha,n_1+n_2-2} \tag{3.6}$$

is used. This is the one-sided t-test. *Balanced samples* are those in which the candidates have an equal number of observations ($n = n_1 = n_2$). These will be considered next.

3.1.3 The Paired t-Test

To compare different run configurations, CRN have been used in our experiments. The reader is referred to Law & Kelton (2000) for a discussion of CRN and related variance-reducing techniques. The jth paired difference

$$d_j = y_{1j} - y_{2j} \qquad j = 1, \dots, n,$$

is used to define the test statistic

$$t_0 = \frac{\overline{d}}{S_d/\sqrt{n}},$$

where $\overline{d} = \frac{1}{n}\sum_{j=1}^{n} d_j$, and

$$S_d = \sqrt{\sum_{j=1}^{n} \frac{(d_j - \overline{d})^2}{n-1}}, \tag{3.7}$$

is the *sample standard deviation of the differences.* Let

$$\delta = \mu_1 - \mu_2$$

denote the difference in means. The null hypothesis $H : \mu_1 = \mu_2$, or equivalently $H : \delta = 0$, would be not accepted if $t_0 > t_{\alpha,n-1}$. The paired t-test can be advantageous compared to the two-sample t-test due to its noise reduction properties. The confidence interval based on the paired test can be much narrower than the corresponding interval from the two-sample test. The reader is referred to the discussion in Montgomery (2001).

3.2 Monte Carlo Simulations

The statistical approach from Example 2.2 requires the following steps:

1. First, a sampling distribution for a statistic is derived.
2. Then the probability of a sample statistic is determined.

Many sampling distributions rely on statistical assumptions; consider, for example, the assumption that samples are drawn from normal distributions like for the t-distribution. Furthermore, classical techniques often apply asymptotic results under the assumption that the size of the available set of samples is sufficiently large. Monte Carlo simulations can be applied for known population distributions from which the samples are drawn and unknown sampling distributions of the test statistic, for example, the trimmed mean or the interquartile range.

As *bootstrap* methods treat the sample as the population, they can be applied if the sampling distribution is unknown (Efron & Tibshirani 1993). They require a representative sample of the population. Nowadays the bootstrap is considered a standard method in statistics (Mammen & Nandi 2004). It has been successfully applied to solve problems that would be too complicated for classical statistical techniques and in situations where the classical techniques are not valid (Zoubir & Boashash 1998).

Bootstrap

The idea behind the bootstrap is similar to a method that is often applied in practice. Experiments are repeated to improve the estimate of an unknown parameter. If a representative sample is available, the bootstrap randomly reassigns the observations and recomputes the estimate. The bootstrap is a computationally intensive technique. Let $\hat{\theta}$ be the estimate of an unknown parameter θ that has been determined by calculating a statistic T from the sample:

$$\hat{\theta} = T = t(y_1, \ldots, y_n).$$

By sampling with replacement, n_b bootstrap samples can be obtained. The bootstrap replicates of $\hat{\theta}$

$$\hat{\theta}^{*b} = t(y^{*b}), \qquad b = 1, \ldots, n_b,$$

provide an estimate of the distribution of $\hat{\theta}$. The generic bootstrap procedure is described in Fig. 3.1.

We describe the basic bootstrap procedure to determine the observed significance level $\alpha_{\overline{d}}(\delta)$. It can be applied to generate plots of the observed significance, as shown in Fig. 2.5. Note that this procedure requires only two paired and representative samples, y_1 and y_2.

Let $y_1 = (y_{11}, \ldots, y_{1n})^T$ and $y_2 = (y_{21}, \ldots, y_{2n})^T$ denote the random samples, and $d = y_1 - y_2 = (y_{11} - y_{21}, \ldots, y_{1n} - y_{2n})^T$ their difference vector.

Algorithm: Generic Bootstrap

1. Calculate $\hat{\theta}$ from a representative sample $y = (y_1 \ldots, y_n)$.
2. To generate the bootstrap data sets $y^{*b} = (y_1^{*b}, \ldots, y_n^{*b})$ sample with replacement from the original sample.
3. Use the bootstrap sample y^{*b} to determine $\hat{\theta}^{*b}$.
4. Repeat steps 2 and 3 n_b times.
5. Use this estimate of the distribution of $\hat{\theta}$ to obtain the desired parameter, for example the mean.

Fig. 3.1. The generic bootstrap procedure

The procedure to obtain an estimate of the observed significance level $\alpha_{\overline{d}}(\delta)$ for a difference $\delta = d_0$ under the null hypothesis H can be implemented as in the following example:

Example 3.1 (Bootstrap). Let y_1 and y_2 denote two vectors with representative samples from a population. If $a \in \mathbb{R}$ and the vector $y = (y_1, \ldots, y_n)^T \in \mathbb{R}^n$, the scalar–vector addition is defined as

$$a + y = (y_1 + a, \ldots, y_n + a)^T.$$

The bootstrap approach to generate the plots of the observed significance comprises the steps shown in Fig. 3.1. They can be detailed as follows:

1. Determine $d = y_1 - y_2$.
2. Determine $\overline{d} = 1/n \sum_{j=1}^{n}(y_{1j} - y_{2j})$.
3. Determine $d_0 = d - \overline{d}$.
4. Specify the lower bound d_l and the upper bound d_u for the plot.
5. Specify m, the number of points to be plotted in the interval $[d_l, d_u]$.
6. For $i = 1$ to m do:
   (a) Determine $d_i = d_l + i \cdot (d_u - d_l)/m + d_0$.
   (b) Generate n_b bootstrap sample sets d_i^{*b}, $b = 1, \ldots, n_b$ from d_i.
   (c) Determine the n_b mean values $\overline{d}_i^{*b}$.
   (d) Determine n_i, that is, the number of times that $\overline{d}_i^{*b} > \overline{d}$.
   (e) Determine the ratio $r_i = n_i/n_b$.

Finally, the m points $(d_0^{(i)}, r^{(i)})$ are plotted. The ratio r_i corresponds to the observed significance value $\alpha_{\overline{d}}(d_0^{(i)})$. ■

Histograms of the bootstrap replicates as shown in Fig. 3.2 are appropriate tools for examining the distribution of $\hat{\theta}$. Figure 3.3 depicts the result based on the bootstrap. It represents the same situation as shown in Fig. 2.5, without making any assumption on the underlying distribution. As the sample size is increased, i.e., from 50 to 500, the bootstrap and the true curve start to look

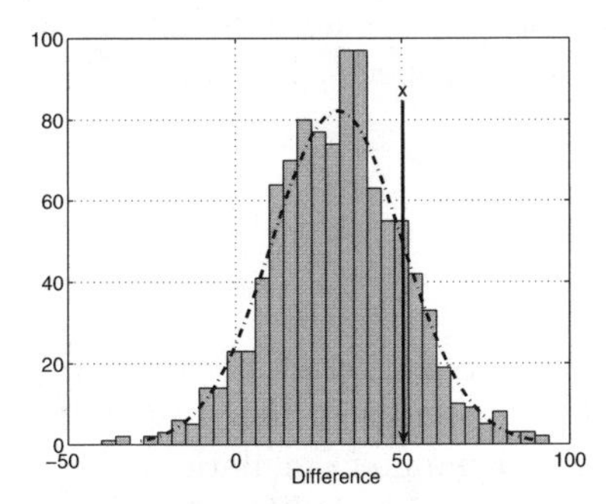

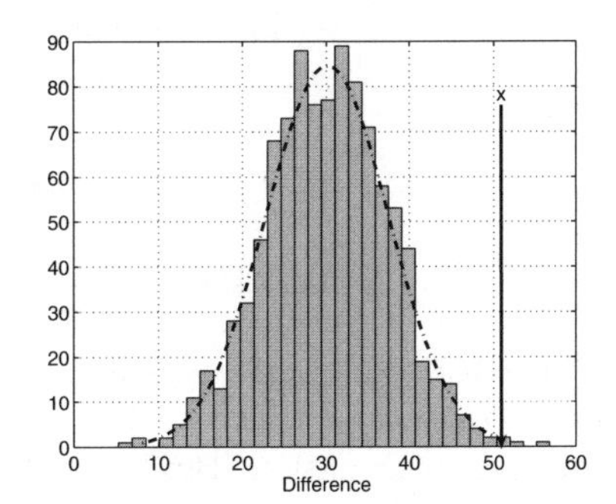

Fig. 3.2. Histograms of the bootstrap samples. *Left*: 50 repeats; *right*: 500 samples. These figures show histograms of the bootstrap samples that were generated at step 6 in Example 3.1. The difference d_i has the value 30. The *dash-dotted curves* show the superimposed normal density. The area to the right of $\overline{d} = 51.73$ under the curve corresponds approximately with the observed significance level $\alpha_{\overline{d}}(\delta)$, the ratio r_i

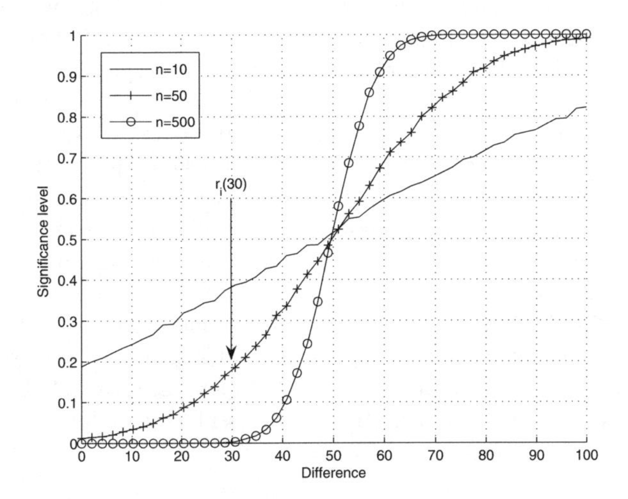

Fig. 3.3. This figure depicts the same situation as shown in Fig. 2.5. But, unlike in Fig. 2.5, no assumptions on the underlying distribution have been made. Samples of size $n = 10$, 50, and 500, respectively, have been drawn from a normal distribution. The bootstrap procedure described in Example 3.1 has been used to generate this plot. The curves look qualitatively similar to the curves from Fig. 2.5. As the number of samples increases, the differences between the exact and the bootstrap curves becomes smaller. The measured difference is 51.73, $\sigma = 160$, cf. Example 2.1. Regard $n = 50$: If the true difference is (a) 0, (b) 51.73, or (c) 100, then (a) $H : \delta = 0$, (b) $H : \delta = 51.73$, or (c) $H : \delta = 100$ is (approximately) wrongly rejected (a) 1%, (b) 50%, or (c) 99% of the time

increasingly similar. The following sections present standard definitions from classical and modern design and analysis of experiments, DOE and DACE, respectively.

3.3 DOE: Standard Definitions

The classical design of experiments has a long tradition in statistics. It has been developed for different areas of applications: agriculture (Fisher 1935), industrial optimization (Box et al. 1978), and computer simulation (Kleijnen 1987). The following definitions are commonly used in DOE. The input parameters and structural assumptions to be varied during the experiment are called *factors* or *design variables*. Other names frequently used are predictor variables, input variables, regressors, or independent variables. The vector of design variables is represented as $x = (x_1, \ldots, x_k)^T$. Different values of parameters are called *levels*. The levels can be scaled to the range from -1 to $+1$. Levels can be qualitative, e.g., selection scheme, or quantitative, e.g., population size. The *design space*, the region of interest, or the experimental region is the k-dimensional space defined by the lower and upper bounds of each design variable. A *sample* or a *design point* is a specific instance of the vector of design variables. An *experimental design* is a procedure for choosing a set of factor level combinations. Kleijnen (2001) defines DOE as "the selection of combinations of factor levels that will be simulated in an experiment with the simulation model." One parameter design setting is run for different pseudorandom number settings, resulting in replicated outputs. The output value y is called *response*; other names frequently used are output variables or dependent variables.

The intuitive definition of a *main effect* of a factor A is the change in the response produced by the change in the level of A averaged over the levels of the other factors. The average difference between the effect of A at the high level of B and the effect of A at the low level of B is called the *interaction* effect AB of factor A and factor B.

3.4 The Analysis of Variance

Following Montgomery (2001), we introduce one of the most useful principles in inferential statistics, the (single) factor *analysis of variance* (ANOVA). First, the dot subscript notation is defined: Consider m different treatments. The sum of all observations under the ith treatment is

$$y_{i\cdot} = \sum_{j=1}^{n} y_{ij}.$$

Then, $\overline{y}_{i\cdot} = y_{i\cdot}/n$, $i = 1, 2, \ldots, m$, and

$$y_{..} = \sum_{i=1}^{m}\sum_{j=1}^{n} y_{ij}, \qquad \overline{y}_{..} = y_{..}/N,$$

where $N = nm$ is the total number of observations. The *total corrected sum of squares*

$$SS_{\mathrm{T}} = \sum_{i=1}^{m}\sum_{j=1}^{n}(y_{ij} - \overline{y}_{..})^2, \tag{3.8}$$

measures the total variability in the data. It can be partitioned into a sum of squares of the difference between the treatment averages and the grand average $\mathrm{SS}_{\mathrm{TREAT}}$ plus a sum of squares of the differences of observations within treatments from the treatment average SS_{E}:

$$\sum_{i=1}^{m}\sum_{j=1}^{n}(y_{ij} - \overline{y}_{..})^2 = n\sum_{i=1}^{m}(\overline{y}_{i.} - \overline{y}_{..})^2 + \sum_{i=1}^{m}\sum_{j=1}^{n}(y_{ij} - \overline{y}_{i.})^2. \tag{3.9}$$

This *fundamental ANOVA principle* can be written symbolically as:

$$SS_{\mathrm{T}} = SS_{\mathrm{TREAT}} + SS_{\mathrm{E}}. \tag{3.10}$$

The term SS_{TREAT} is called the *sum of squares due to the treatments*, and SS_{E} is called the *sum of squares due to error.*

3.5 Linear Regression Models

(Linear) regression models are central elements of the classical design of experiments approach. In stochastic simulation and optimization, regression models can be represented as follows:

$$y = f(z_1, \dots, z_d, r_0), \tag{3.11}$$

where f is a mathematical function, e.g., $f : \mathbb{R}^{d+1} \to \mathbb{R}$: Given the values of the argument z_i and at least one random number seed r_0, the simulation program or the optimization algorithm determine exactly one value. Least square methods can be applied to estimate the *linear model*

$$y = X\beta + \epsilon, \tag{3.12}$$

where y denotes a column vector with the N responses, ϵ is the vector of N error terms, and β denotes the vector with q parameters β_j ($N \geq q$). Usually, the normality assumption (the error term ϵ is normal with *expectation* $E(\epsilon) = 0$ and *variance* $V(\epsilon) = \sigma^2$) is made. The *regression matrix* X is the $(N \times q)$ matrix:

$$X = \begin{pmatrix} 1 & x_{11} & x_{12} & \cdots & x_{1,q-1} \\ \vdots & & & & \vdots \\ 1 & x_{i1} & x_{i2} & \cdots & x_{i,q-1} \\ \vdots & & & & \vdots \\ 1 & x_{N1} & x_{N2} & \cdots & x_{N,q-1} \end{pmatrix}. \tag{3.13}$$

Let $\mathbf{1}$ denote the vector of ones: $\mathbf{1} = (1, 1, \ldots, 1)^T$. The variable x_0 is the dummy variable equal to $\mathbf{1}$, and the remaining $q - 1$ variables may correspond to the simulation or optimization parameters z_i. Let I_n denote the n-dimensional identity matrix. Experimental settings (designs), where the regression matrix X satisfies $X^T X = NI_q$, are called orthogonal. The *ordinary least squares* estimator of the vector of regression parameters β in Eq. (3.12) reads:

$$\hat{\beta} = (X^T X)^{-1} X^T y \tag{3.14}$$

with covariance

$$\text{cov}(\hat{\beta}) = \sigma^2 (X^T X)^{-1}. \tag{3.15}$$

An estimator $\hat{\alpha}$ is unbiased if $E(\hat{\alpha}) = \alpha$. If the errors in Eq.(3.12) are independently and identically distributed, then $\hat{\beta}$ is the *best linear unbiased estimator*. An example how to apply regression models to analyze the performance of evolutionary algorithms is given in Bartz-Beielstein (2003).

Generalized Linear Models

Linear models are applicable to problems that have Gaussian errors. In many situations the optimization practitioner has to face response values that follow some skewed distribution or have nonconstant variance. To deal with nonnormal responses, data transformations are often recommended, although the choice of an adequate transformation can be difficult. Draper and Smith (1998) discuss the need for transformation and present different transformation methods. Since the transformation may result in incorrect values for the response value, i.e., $\log Y$, if $Y < 0$, GLMs provide an alternative (McCullagh & Nelder 1989). François & Lavergne (2001) and Bartz-Beielstein (2003) use GLMs to analyze evolutionary algorithms. Bartz-Beielstein et al. (2005c) propose GLMs to analyze and validate simulation models.

Logistic regression models that are based on the *success ratio* (SCR) can be used to analyze the algorithm's performance. Whether or not the optimization run has located a prespecified optimum can be used as a performance measure for algorithms. In this case, where the outcome variable can take only two values, a linear regression model is not appropriate, but a logistic regression model might be adequate. The number of successful runs can be seen as a random variable having a binomial distribution. For an introduction into logistic regression, the reader is referred to Collett (1991). Myers & Hancock (2001) present an example that uses a genetic algorithm to solve consistent labeling problems.

Standard textbooks on regression analysis such as Draper & Smith (1998) present methods of checking the fitted regression model. However, the fact that the regression model passes some test does not mean that it is the correct model. Graphical tools should be used to guide the analysis. Half-normal plots, design plots, interaction plots, box plots, scatter plots, and trellis plots that can be applied to analyze computer experiments will be presented next.

3.6 Graphical Tools

This section presents graphical tools that support the analysis of factor effects and interactions. Half-normal plots, interaction plots, and box plots can complement classical DOE methods. They are based on factorial designs (designs will be introduced in Chap. 5). Scatter plots can be used in combination with space-filling designs. These designs are commonly used in modern DACE.

3.6.1 Half-Normal Plots

Least-squares estimation gives an estimate of the effect of a factor. The estimated effects minimize the sum of squared differences between raw data and the fitted values from the estimates. An ordered list of the main effects (and of the interactions as well) can be constructed. A half-normal plot is a plot of the absolute value of the effect estimates against their cumulative normal probabilities.

Example 3.2. Figure 3.4 depicts a typical half-normal plot that has been generated while optimizing an evolution strategy (Mehnen et al. 2004a). Evolution strategies and other algorithms are introduced in Chap. 6. Regarding this half-normal plot, τ_0^m, the multiplier for the individual and global mutation parameters, the population size μ, and the selection pressure have a significant influence on the algorithm's performance. In addition, the interactions between μ and τ_0^m, and μ and ν play a significant role. ■

3.6.2 Design Plots

Design plots visualize the means of each factor and plot $100(1-\alpha)$ confidence intervals around each mean. Design plots should be complemented with interaction plots. They might be misleading, because they are intuitively appealing and easy to "understand." Figure 8.6 gives an illustrative example.

3.6.3 Interaction Plots

If the number of factor levels is low, e.g., in two-factor experiments, *interaction plots* can be used to check for interactions between the factors. Interaction plots show how pairs of factors, i.e., selection method and population size in evolution strategies, interact in influencing the response (Y).

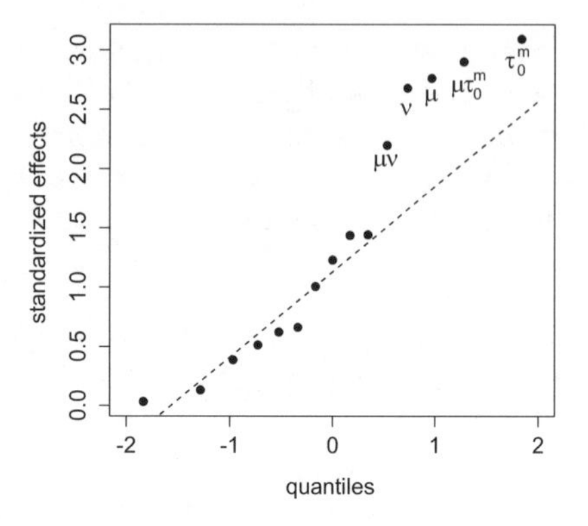

Fig. 3.4. Half-normal plot to detect important factors of an evolution strategy. The multiplier for the individual and global mutation parameters τ_0^m, the population size μ, and the selection pressure have a statistically significant influence on the algorithm's performance. In addition, the interactions between μ and τ_0^m, and μ and ν play a significant role (Mehnen et al. 2004a)

Example 3.3. Consider Fig. 3.5. The horizontal axes of the plots show levels of the first factor: comma selection (Comma), plus selection (Plus), and *threshold selection* (TS). Lines are drawn for the mean of the response for the corresponding level of the interaction between selection method and selective strength (left panel) and selection method and population size (right panel). As the lines run in parallel in both panels, no interactions, which might make the analysis difficult, can be detected. These figures indicate that the selection method TS improves the performance independently from the population size or selective pressure (Beielstein & Markon 2001). ■

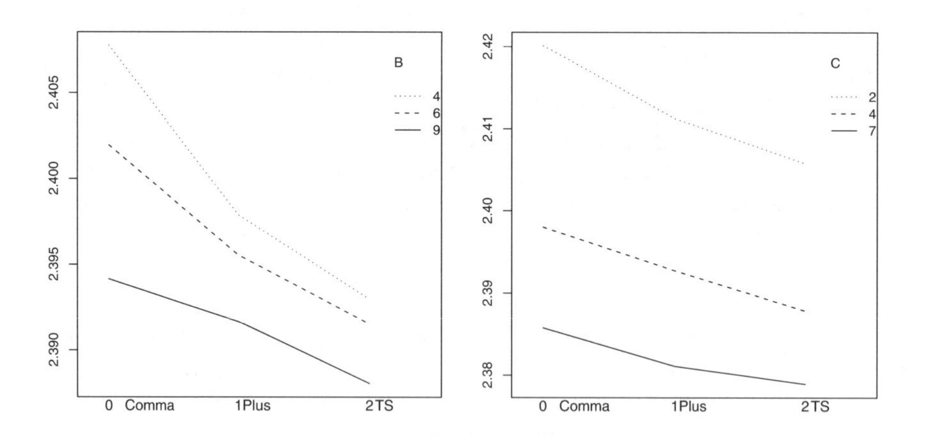

Fig. 3.5. Interaction plots. Plot of the means of the responses. The labels on the x-axis represent different selection mechanisms for an evolution strategy: comma selection (0), plus selection (1), and threshold selection (2). Selective strength (B) with levels 4, 6, and 9 and population size (C) with levels 2, 4, 7 have been chosen for the comparison (Beielstein & Markon 2001)

3.6.4 Box Plots

Box plots as illustrated in Fig. 3.6 display the distribution of a sample. They are excellent tools for detecting changes between different groups of data (Chambers et al. 1983). Let the *interquartile range* (IQR) be defined as the difference between the first and the third sample quartiles. Besides the three sample quartiles (the lower, the median, and the upper quartiles), the minimum and the maximum values, two limits are used to generate the box plots: $y_l = q_{.25} - 1.5\text{IQR}$ and $y_u = q_{.25} + 1.5\text{IQR}$. Possible outliers may lie outside the interval $[y_l, y_u]$. Figure 7.11 shows an example.

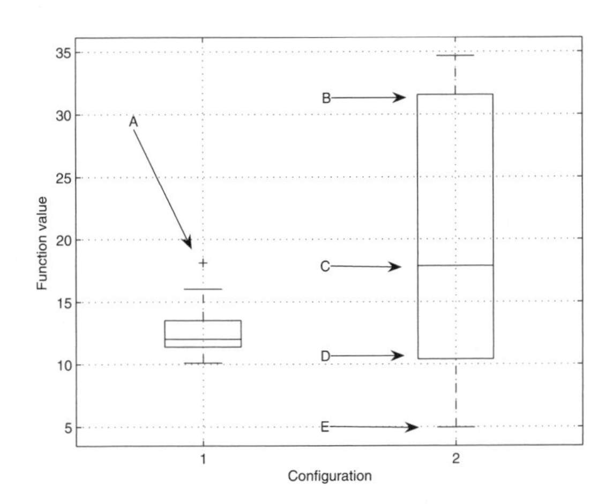

Fig. 3.6. Five elements of a box plot. This figure shows possible outlier A, quartiles B, C, D, and adjacent value E

3.6.5 Scatter Plots

A *scatter plot* is a simple way to visualize data (Chambers et al. 1983; Croarkin & Tobias 2004). It displays important information on how the data are distributed or the variables are related: Are the design variables x_1 and x_2 related? Are variables x_1 and x_2 linearly or nonlinearly related? Does the variation in x_1 change depending on x_2? Are there outliers? This information should be taken into account before any statistical model is built. Scatter plots have been used to detect factor settings that produced outliers and to determine suitable variable ranges.

Example 3.4. Each panel in Fig. 3.7 depicts a scatter plot of the response against one factor. The relationship between the function values and different levels of social parameter c_2 of a particle swarm optimization is shown in the lower right-hand panel. Settings with $c_2 < 2.5$ produced many outliers. Reducing the region of interest for this variable from $[0, 4]$ to $[2.5, 4]$ resulted in

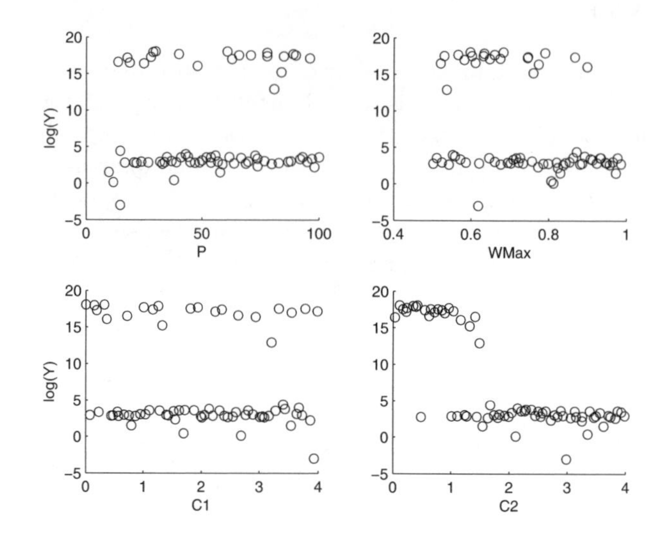

Fig. 3.7. Scatter plots of the response $\log(y)$ plotted for different values of the exogenous strategy parameters of a particle swarm optimization (p, c_1, c_2, and $w_{\max}$) while optimizing Rosenbrock's banana function. A Latin hypercube design was used to generate the data. These plots indicate that modifying the variable range of the social parameter c_2 from $[0, 4]$ to $[2.5, 4]$ leads to improved results (fewer outliers)

fewer outliers. Similar results could have been obtained with box plots, design plots, or other tools from exploratory data analysis. No high-level statistical model assumptions are necessary to perform this analysis. ■

3.6.6 Trellis Plots

Trellis plots depict relationships between different factors through conditioning. They show how plots of two factors change with variations in a third, the so called conditioning factor. Trellis plots consist of a series of panels where each panel represents a subset of the complete data divided into subintervals of the conditioning variable.

Example 3.5. Two variation operators DES and ES, and several population sizes were analyzed. The data points have been divided into four intervals I_1–I_4 due to their population-size values (Fig. 3.8): $I_1 = [7.5, 12.5]$ with 11 data points, $I_2 = [7.5, 17.5]$ with 16 data points, $I_3 = [12.5, 22.5]$ with 16 data points, and $I_4 = [17.5, 22.5]$ with 11 data points.

Fig. 3.8 indicates that ES variation performs significantly better than DES variation. The trellis plots show that this effect occurs independently from the settings of the population size, i.e., there is no interaction between these factors. ■

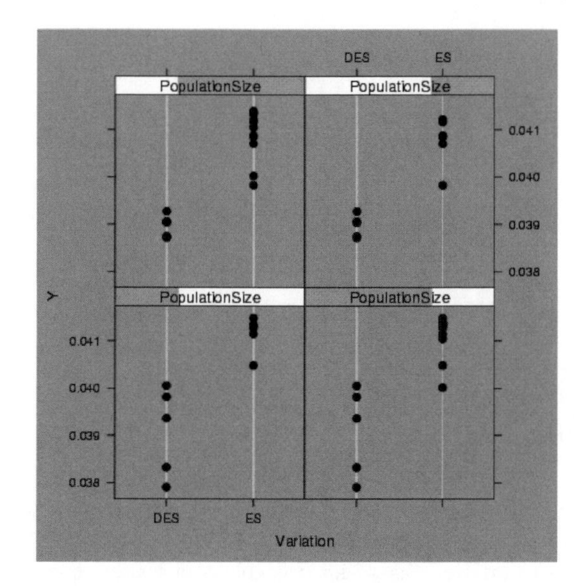

Fig. 3.8. Trellis plots. Algorithm's performance measured as hypervolume Y. These figures support the assumption that ES variation outperforms DES variation significantly. Note, larger hypervolume values (Y) are better in this graph

There are many other useful visualization techniques. To illustrate the online behavior of algorithms, plots of the best function values, of positions in the search space, of endogenous parameter settings (e.g., step sizes in evolution strategies), or diversity plots are frequently generated. Off-line visualization techniques model data after the run is finished. They can be used to plot successful and unsuccessful starting points (related to the success rate) and other performance measures that will be discussed in Chap. 7.

3.7 Tree-Based Methods

Van Breedam (1995) applied tree-based classification methods to analyze algorithms. He used an *automatic interaction detection* (AID) technique developed by Morgan & Sonquist (1963) to determine the significant parameter settings of genetic algorithms. Breiman et al. (1984) introduced *classification and regression trees* (CART) as a "flexible nonparametric tool to the data analyst's arsenal." Tree-based methods can be deployed for screening variables and for checking the adequacy of regression models (Therneau & Atkinson 1997). AID and CART use different pruning and estimation techniques.

The construction of regression trees can be seen as a type of variable selection (Chambers & Hastie 1992; Hastie et al. 2001). Consider a set of design variables $X = \{x_1, \ldots, x_d\}$ and a quantitative response variable Y. Design variables are called *predictor variables* in the context of CART. A regression tree is a collection of rules such as "if $x_1 \leq 5$ and $x_4 \in \{A, C\}$, then the predicted value of Y is 14.2," which are arranged in a form of a binary tree (see, e.g., Fig. 3.9).

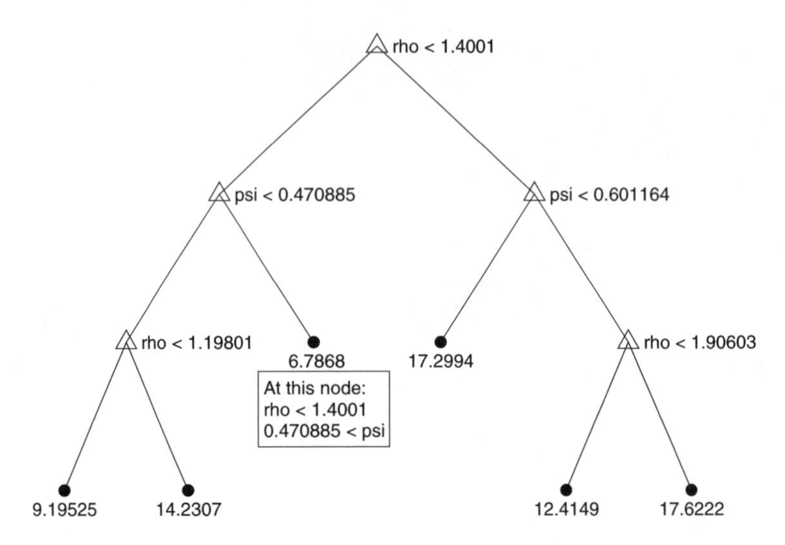

Fig. 3.9. Pruned regression tree to analyze the design variable of a Nelder–Mead simplex algorithm. The left subtree of a node contains the configurations that fulfill the condition in the node. It is easy to see that smaller ρ and larger ψ values improve the algorithm's performance

The binary tree is built up by recursively splitting the data in each node. The tree can be read as follows: If the rule that is specified at the node is true, then take the branch to the left; otherwise take the branch to the right. The partitioning algorithm stops when the node is homogeneous or the node contains too few observations. If qualitative and quantitative design variables are in the model, then *tree-based models* are easier to interpret than linear models. The endpoint of a tree is a partition of the space of possible observations.

Tree construction (TC) comprises three phases (Martinez & Martinez 2002):

(TC-1) *Growing.* In the first phase of the construction of a regression tree a large tree $T_{\max}$ is grown. The partitioning procedure requires the specification of four elements: a splitting criterion, a summary statistic to describe a node, the error of a node, and the prediction error (Therneau & Atkinson 1997). The splitting process can be stopped when a minimum node size is reached. Consider a *node* v. A *leaf* l is any node that has no child nodes, and T_L denotes the *set of all leaves* of a tree T. A *subtree* is the tree which is a child of a node.

The summary statistic is given by the *mean of the node* $\overline{y}(v)$, which is defined as the average response of the cases that fulfill the condition in the node:

$$\overline{y}(v) = \frac{1}{n_v} \sum_{x_i \in v} y_i,$$

where n_v denotes the number of cases in this node. The *squared error of a node* is related to the variance of $y(v)$. It is defined as

$$R(v) = \frac{1}{n} \sum_{x_i \in v} [y_i - \overline{y}(v)]^2 ,$$

where n denotes the size of the entire sample. The *mean squared error for the tree* T is obtained by adding up all of the squared errors in all of the leaves:

$$R(T) = \sum_{l \in T_L} R(l).$$

The mean squared error for the tree is also referred to as the *total within-node sum of squares.* As a splitting criterion, the change in the mean squared error for a split s_v is used:

$$\Delta R(s_v) = R(v) - (R(v_L) + R(v_R)),$$

where v_L and v_R denote the left and right subtrees with root node v, respectively. The best split s_v^* is the split that maximizes the change in the mean squared error $\Delta R(s_v)$.

(TC-2) *Pruning.* The large tree $T_{\max}$ is pruned back in a second phase of the tree construction. The pruning procedure uses a cost-complexity measure:

$$R_{c_p}(T) = R(T) + c_p n_L, \quad c_p \geq 0,$$

where n_L is the number of leaves. As a large tree with leaves that contain only cases from one class has a mean squared error $R(T) = 0$, the c_p value represents the *complexity cost* per leaf. Take the tree that minimizes $R_{c_p}(T)$ to be optimal for the given value of c_p. Note that the value of $R(T)$ decreases as the size of the tree is increased, while $c_p n_L$ increases. By increasing c_p, we can move from one optimum to the next. Each move might require a reduced tree, because the optimal tree size decreases as c_p increases. The pruning procedure constructs a finite sequence of optimal subtrees such that

$$T_{\max} > T_1 > T_2 > \ldots > T_k > \ldots > T_K = \{v_1\},$$

where $\{v_1\}$ is the root node of the tree, and

$$0 = c_{p_1} < \ldots < c_{p_k} < c_{p_{k+1}} < \ldots < c_{p_K}.$$

(TC-3) *Selection.* Finally, the "best" tree is chosen in the selection phase from the sequence of subtrees generated in step TC-2. To select the right tree, the *cross-validation estimate for the prediction error* $R_{\text{CV}}(T_k)$ for each tree in the sequence of pruned trees and the associated estimate of the *standard error of the cross-validation estimate of the prediction error* $s_R(T_k)$ are determined next. Let T' denote the subtree that has

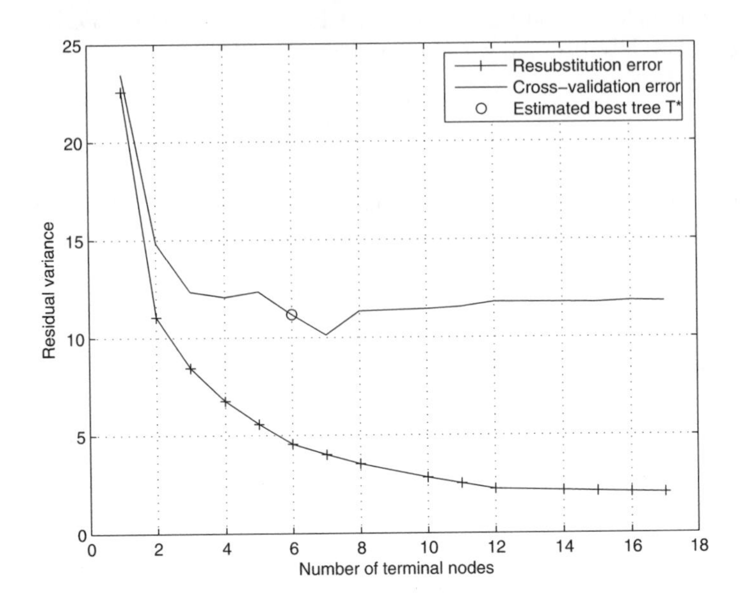

Fig. 3.10. Visualization of the 1-SE (one standard error) selection rule to determine the right tree. The tree with the smallest cross-validation error R_{CV} is not chosen. The 1-SE rule chooses the tree with six nodes, because its error lies in the 1-SE corridor of the tree with the smallest error (the tree with seven nodes). The pruned tree with six nodes is shown in Fig. 3.9. The full tree has 17 nodes. In addition, the resubstitution error is plotted against the number of nodes in the tree. It is monotonically decreasing because it does not measure the costs for including additional nodes in the tree

the smallest estimated prediction error. Its standard error is denoted as $s_R(T')$. The *one standard error rule* (1-SE rule) selects the smallest tree T^* with

$$R_{\text{CV}}(T^*) \leq R_{\text{CV}}(T') + s_R(T'). \tag{3.16}$$

The tree selection procedure is depicted in Fig. 3.10. The *resubstitution error*, which is calculated from the whole data set, is also shown. It should only be used to control the tree selection procedure because it gives an optimistic assessment of the relative error.

The tree with the smallest error is not chosen, but instead the smallest tree that reaches the error-corridor of the smallest error plus one standard error (1-SE rule). The 1-SE rule chooses the tree with six nodes, since its error lies in the 1-SE corridor of the tree with seven nodes.

3.8 Design and Analysis of Computer Experiments

We consider each algorithm run as a realization of a stochastic process. *Kriging* is an interpolation method to predict unknown values of a stochastic process and can be applied to interpolate observations from computationally expensive simulations (Isaaks & Srivastava 1989). Our presentation follows concepts introduced in Sacks et al. (1989), Jones et al. (1998), and Lophaven et al. (2002b).

3.8.1 The Stochastic Process Model

The regression model

$$y = X\beta + \epsilon,$$

cf. Eq. (3.12), requires that the response y and the error ϵ have the same variance. The assumption of a constant variance is unrealistic in many simulation and optimization scenarios. A variance that varies with x appears to be more realistic. For example, Kleijnen (1987) reports that the standard errors of the y_i's in simulation models differ greatly. Similar ideas are presented in Jones et al. (1998), where stochastic process models as alternatives to regression models are introduced.

Consider a set of m design points $x = (x_1, \ldots, x_m)^T$ with $x_i \in \mathbb{R}^d$ as in Sect. 3.5. In the *DACE stochastic process model*, a deterministic function is evaluated at the m design points x. The vector of the m responses is denoted as $y = (y_1, \ldots, y_m)^T$, with $y_i \in \mathbb{R}$. The process model proposed in Sacks et al. (1989) expresses the deterministic response $y(x_i)$ for a d-dimensional input x_i as a realization of a regression model $\mathcal{F}$ and a stochastic process Z:

$$Y(x) = \mathcal{F}(\beta, x) + Z(x). \tag{3.17}$$

3.8.2 Regression Models

We use q functions $f_j : \mathbb{R}^d \to \mathbb{R}$ to define the regression model

$$\mathcal{F}(\beta, x) = \sum_{j=1}^{q} \beta_j f_j(x) = f(x)^T \beta.$$

Regression models with polynomials of orders 0, 1, and 2 have been used in our experiments. The constant regression model with $q = 1$ reads $f_1(x) = 1$; the linear model with $q = d + 1$ is $f_1(x) = 1, f_2(x) = x_1, \ldots, f_{d+1}(x) = x_d$; and the quadratic model: $f_1(x) = 1$, $f_2(x) = x_1, \ldots, f_{d+1}(x) = x_d$, $f_{d+2}(x) = x_1 x_1, \ldots, f_{2d+1}(x) = x_1 x_d$, …, $f_q(x) = x_d x_d$.

Figure 3.11 illustrates the difference between regression based on models of order 0 and order 1.

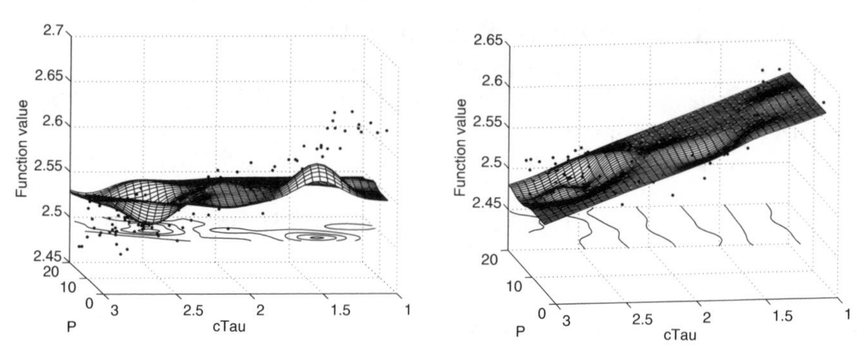

Fig. 3.11. Joint effects of population size P and learning rate $cTau$ of an evolution strategy in the DACE model. *Left*: regression model of order 0; *right*: the same data model with a regression model of order 2. A Gaussian correlation model was used for both cases

3.8.3 Correlation Models

The random process $Z(\cdot)$ is assumed to have mean zero and covariance $V(w, x) = \sigma^2 \mathcal{R}(\theta, w, x)$ with process variance σ^2 and correlation model $\mathcal{R}(\theta, w, x)$. Correlations of the form

$$\mathcal{R}(\theta, w, x) = \prod_{j=1}^{d} \mathcal{R}_j(\theta, w_j - x_j)$$

will be used in our experiments. The correlation function should be chosen with respect to the underlying process (Isaaks & Srivastava 1989). Lophaven et al. (2002a) discuss seven different models. Well-known examples are

$$\begin{aligned}
\text{EXP}: &\quad \mathcal{R}_j(\theta, h_j) = \exp(-\theta_j |h_j|), \\
\text{EXPG}: &\quad \mathcal{R}_j(\theta, h_j) = \exp(-\theta_j |h_j|^{\theta_{d+1}}), \quad 0 < \theta_{d+1} \le 2, \\
\text{GAUSS}: &\quad \mathcal{R}_j(\theta, h_j) = \exp(-\theta_j h_j^2),
\end{aligned} \tag{3.18}$$

with $h_j = w_j - x_j$, and for $\theta_j > 0$. The *exponential correlation function* EXP and the *Gaussian correlation function* GAUSS have a linear and a parabolic behavior, respectively, near the origin. The *general exponential correlation function* EXPG can have both shapes. Large θ_j's indicate that variable j is active: function values at points in the vicinity of a point are correlated with Y at that point, whereas small θ_j's indicate that also distant data points influence the prediction at that point.

Maximum likelihood estimation methods to estimate the parameters θ_j of the correlation functions from Eq. (3.18) are discussed in Lophaven et al. (2002a).

DACE methods provide an estimation of the prediction error on an untried point x, the *mean squared error* (MSE) of the predictor

$$\mathrm{MSE}(x) = E\left(\hat{y}(x) - y(x)\right). \tag{3.19}$$

3.8.4 Effects and Interactions in the Stochastic Process Model

Santner et al. (2003) recommend to use a small design to determine important factor levels. After running the optimization algorithm, scatter plots of each input versus the output can be analyzed. Welch et al. (1992) advocate the use of sensitivity analysis. A screening algorithm that is similar in spirit to forward selection in classical regression analysis is used to identify important factors. Sacks et al. (1989) propose an ANOVA-type decomposition of the response into an average, main effects for each factor, two-factor interactions, and higher-order interactions.

Example 3.6 (Effects in the DACE model). Let the average of $y(x)$ over the experimental region be

$$\mu_0 = \int y(x) \prod_h \mathrm{d}x_h.$$

Define the main effect of factor x_i averaged over the other factors by

$$\mu_i(x_i) = \int y(x) \prod_{h \neq i} \mathrm{d}x_h - \mu_0,$$

and the interaction effect of x_i and x_j by

$$\mu_{ij}(x_i, x_j) = \int y(x) \prod_{h \neq i,j} \mathrm{d}x_h - \mu_0 - \mu_i(x_i) - \mu_j(x_j).$$

Higher-order interactions can be obtained accordingly. To estimate these effects, $y(x)$ is replaced by $\hat{y}(x)$. Factors for which the predicted response is not sensitive can be fixed in subsequent modeling steps. A similar approach was proposed by Schonlau (1997) to plot the estimated effects of a subset x_{effect} of the x variables.

Figure 3.12 illustrates the main effects in the DACE model that has been generated while optimizing the exogenous strategy parameters of a (1 + 1)-ES. See Sect. 7.5 for the discussion of this sequential parameter optimization process. The four panels on the left show the estimated effects that have been predicted with the first stochastic process model. By adding further design points the model is improved. Predictions based on the improved model are shown in the four panels on the right. ■

As the design plots should be complemented with interaction plots, we recommend to complement the *effect plots* from Fig. 3.12 with interaction plots that can be produced with the DACE toolbox (Lophaven et al. 2002b). These three-dimensional visualizations can be used to illustrate the interaction between two design variables as well as the relationship between these design variables and the associated mean squared error of the predictor from Eq. (3.19).

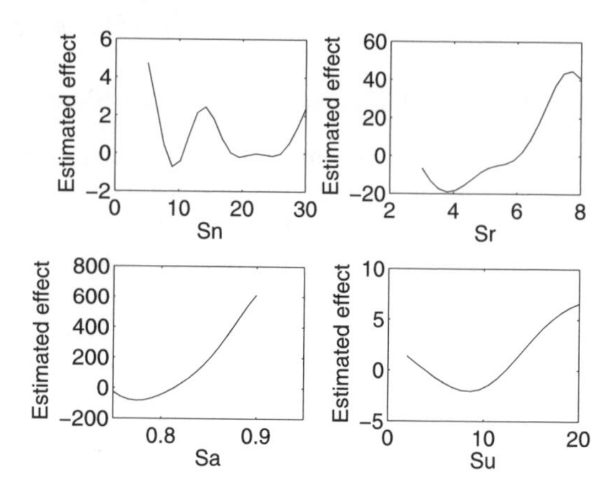

Fig. 3.12. Effects in the DACE model. The model was based on data from case study I (Sect. 8.2)

3.9 Comparison

Comparing classical linear regression models and tree-based regression models, we can conclude that regression trees present results in an intuitively understandable form. The results are immediately applicable, and interactions are automatically included. Regression trees can handle qualitative and quantitative factors. On the other hand, it is not guaranteed that the overall tree is optimal. The splitting criteria are only locally optimal; that is, it is only guaranteed that each single split will be optimal. Trees can become quite large and can then make poor intuitive sense. If only a small set of data is available, parametric models might be advantageous that are based on strong assumptions regarding the underlying distribution or the form of the linear model.

To compare different regression techniques the following criteria might be useful: Is the technique flexible; that is, can it cope with different types of variables (quantitative, qualitative) and does not require assumptions on the underlying distribution? Are the results plausible, can even complex interactions be detected? Is the method able to model gathered data and to predict new values? The availability of related literature and software packages should be judged as well.

A combination of different techniques is useful. Tree-based techniques may produce different insights than regression models. Regression trees can be used at the first stage to screen out the important factors. If only a few quantitative factors remain in the model, DACE techniques can be applied to get an exact approximation of the functional relationship between parameter settings and an algorithm's performance. Sequential designs have been applied successfully during our analysis (Bartz-Beielstein & Markon 2004; Bartz-Beielstein et al. 2004b).

3.10 Summary

The basic ideas from statistics presented in this chapter can be summarized as follows:

1. A statistical hypothesis H is a statement regarding the parameters of a distribution.
2. Hypothesis testing is the procedure of formulating a hypothesis H, taking a random sample, computing a test statistic, and the acceptance of (or failure to accept) H.
3. Paired data can simplify the statistical analysis.
4. Computer-intensive methods enable the estimation of unknown parameters.
   (a) Monte Carlo simulations can be used to estimate unknown parameters of a known distribution.
   (b) The bootstrap requires a representative sample from the population only.
   (c) It can be used to generate plots of the observed significance if the sample distribution is unknown.
5. Standard definitions from DOE:
   (a) Factors are input parameters to be varied during the experiment. Their different values are called levels. Factors can be qualitative or quantitative.
   (b) Upper and lower bounds of each factor specify the region of interest.
   (c) Output values are called responses.
   (d) The main effect of a factor is the change in the response produced by the change in the level of this factor averaged over the levels of the other factors.
   (e) An experimental design comprises a problem design and an algorithm design.
6. The total variability in the data can be partitioned into the variability between different treatments and the variability within treatments:
$$SS_T = SS_{\text{TREAT}} + SS_{\text{E}}.$$
   This fundamental ANOVA principle is used in DOE, tree-based regression, and DACE.
7. Linear regression models are based on the equation $y = X\beta + \epsilon$, where X denotes a regression matrix that represents the experimental design, ϵ is a vector of error terms, y is a vector with responses, and β is a vector that represents the model parameters.
8. Half-normal plots, design plots, scatter plots, interaction plots, trellis plots, and box plots are tools from exploratory data analysis to visualize the distribution of the data and possible relations between factors.
9. The three basic steps to construct a regression tree are growing, pruning, and selecting. Tree-based methods cannot replace, but they should complement classical methods.

10. The assumption of homogeneous variance in the regression model $y = X\beta + \epsilon$ (Eq. (3.12)) appears to be unrealistic. DACE models include this inhomogeneity in their model specification.
11. The DACE stochastic process model expresses the deterministic response as a realization of a regression model and a random process.
12. An ANOVA-like decomposition can be used in DACE to visualize the factor and interaction effects.

3.11 Further Reading

Montgomery (2001) presents an introduction to the classical design of experiments, the bootstrap is introduced in Efron & Tibshirani (1993), classification and regression trees are discussed in Breiman et al. (1984), and Santner et al. (2003) describe the design and analysis of computer experiments. The MATLAB toolbox DACE provided by Lophaven et al. (2002b) implements the basic functions to construct stochastic process models.

4

Optimization Problems

> Don't get involved in partial problems, but always take flight to where there is a free view over the whole single great problem, even if this view is still not a clear one.
> —Ludwig Wittgenstein

A well-established experimental procedure in evolutionary computation and related scientific disciplines, like operations research or numerical analysis, to judge the performance of algorithms can be described as shown in Fig. 4.1. This framework relies on the assumption that the experimenter can find the

Heuristic: Comparing Algorithms

1. Define a set of test functions (and an associated testing environment to specify starting points, termination criteria, etc.).
2. Perform tests.
3. Report the performances of the algorithms, for example the number of successfully completed runs. Obviously, the algorithm with the highest (expected) performance is considered best.

Fig. 4.1. Heuristic for the comparison of two algorithms

best algorithm out of a set of potential candidate algorithms. *Test suites* are commonly used to compare the performance of different optimization algorithms. We assert that results from test suites provide useful means to validate whether the implementation of an algorithm is correct (validation). They can be seen as a starting point for further investigations. Why results from test suites can give no satisfying answers to the research goals RG 2.1–2.4 introduced in Sect. 2.1.2 will be discussed in this chapter. We will consider

1. test functions
2. real-world optimization problems
3. randomly generated test problems

Solutions for unconstrained, global optimization problems are defined as follows. Consider a real-valued function $f : \mathbb{R}^d \rightarrow \mathbb{R}$. A *local minimizer* is a *point*

x^* such that there exists an ϵ environment $U_\epsilon(x^*)$ of x^* ($\epsilon > 0$) with

$$f(x^*) \leq f(x), \quad \forall\, x \in U_\epsilon(x^*). \tag{4.1}$$

The related *minimization problem* is written as $\min_x f(x)$. We will consider unconstrained minimization problems only. If $f(x^*) \leq f(x)$ holds for all $x \in \mathbb{R}^d$, x^* is called a *global minimizer*. The symbol $f(x)$ denotes the (objective) *function value* at x. The following section presents problems related to test functions discussed in the EC community.

4.1 Problems Related to Test Suites

Even in the well-structured world of mathematical optimization, a reasonable choice of test functions to evaluate the effectiveness and the efficiency of optimization algorithms is not trivial. However, even if the selection of a test problem is assumed to be unproblematic, the choice of specific problem instances can cause additional problems (a problem instance is a realization of a generic optimization problem). Hence, for a given problem many different instances can be considered. For example, varying the dimension defines different instances of the sphere function. Here we can mention two important difficulties:

- The distribution of instances might influence the algorithm's performance significantly. Although the result from Goldberg (1979) and Goldberg et al. (1982) suggested that the propositional *satisfiability problem* (SAT) can be solved on average in $O(n^2)$, Franco & Paull (1983) showed that this result was based on a distribution of instances "with a preponderance of easy instances" (Mitchell et al. 1992). Neither was the algorithm clever, nor the problem easy: Goldberg sampled from the space of all problem instances without producing any hard cases.
- Increasing the dimensionality of a problem can make a test function easier. This was demonstrated for Griewangk's function, because the number of local optima decreases in number and complexity as the dimensionality increases (Whitley et al. 1996).

Whitley et al. (1996) discuss further aspects of test functions, i.e., symmetry and separability. A test function is *separable* if the global optimum can be located by optimizing each variable independently. A two-dimensional function is *symmetric* if $f(x, y) = f(y, x)\ \forall x, y \in \mathbb{R}^d$. They state that "Surprisingly, almost all of the functions in current evolutionary search test suites are separable." Therefore, they propose new test functions. One of these functions, the `whit` function, is listed in Table 4.1.

A generic test suite might lead to algorithms that perform well on this particular test suite only. The recommendation of many authors to define heterogeneous test suites is merely an apparent solution. To avoid useless and misleading results, it is important to understand why an algorithm performs well or not so well.

It is common practice to finish an article with presenting tabularized result data. The raw data from these tables require a correct interpretation and should not be seen as final results but as starting points for interpretation.

It can be observed that the performance of optimization algorithms crucially depends on the *starting point* $x^{(0)}$ and other start conditions. To put more emphasis on testing the robustness (effectivity), Hillstrom (1977) proposed using *random starting points*. However, random starting points may cause new difficulties, see the discussion in Sect. 5.5.

The *no free lunch theorem* (NFL) for search states that there does not exist any algorithm that is better than another over all possible instances of optimization problems. However, this result does not imply that we should not compare different algorithms. Keeping in mind that we are considering problems of practical interest, the reader may be referred to the discussions in Whitley et al. (1995), Droste et al. (2000), and Whitley et al. (2002).

The problems presented in this subsection can be solved, for example, by building better test functions. But there are other, more severe objections against the concept of strategy comparison stated in Fig. 4.1, as will be seen in Chap. 7.

4.2 Test Functions

Some test functions have become very popular in the EC community. Table 4.1 lists some common functions for global, unconstrained optimization.

To differentiate between test functions for efficiency and effectivity (robustness), Schwefel (1975) proposed three test scenarios: The first tests were performed to analyze the rates of convergence for quadratic objective functions, and the second series to test the reliability of convergence for the general nonlinear case. In a third test, the dependency of the computational effort on the problem dimension for nonquadratic problems was studied. Therefore, problem dimensions from 3 to 1000 have been used. The problem dimensions of the second scenario were relatively small.

4.2.1 Test Function for Schwefel's Scenario 1 and 2

The following function has been used in test series one and two, see also Table 4.1:

(Sphere) Minimum $x_i^* = 0$, for $i = 1, \dots, d$. Optimum $f^* = 0$.

4.2.2 Test Functions for Schwefel's Scenario 2

The Rosenbrock function and the Schwefel function have been used in the second test scenario (Rosenbrock 1960; Schwefel 1975).

Table 4.1. Common test functions based on Whitley et al. (1996). The reader is referred to Schwefel (1995) for a more detailed discussion of test functions. Test problem instances from the S-ring optimization problem are presented in Table 4.3

Symbol	Name	Function
`sphere`:	sphere	$\sum_{i=1}^d x_i^2$
`rosen`:	Rosenbrock	$100(x_1^2 - x_2)^2 + (1 - x_1)^2$
`step`:	step	$\sum_{i=1}^d \lfloor x_i \rfloor$
`quartic`:	quartic function with noise	$\left(\sum_i^d i x_i^4\right) + \mathcal{N}(0,1)$
`shekel`:	Shekel's foxholes	$\left(0.002 + \sum_{j=1}^{25} 1/(j + \sum_{i=1}^2 (x_i - a_{ij})^6)\right)^{-1}$
`rast`:	Rastrigin	$10d\left(\sum_{i=1}^d \left(x_i^2 - 10\cos(2\pi x_i)\right)\right)$
`schwe`:	Schwefel	$-x \sin\left(\sqrt{\lvert x \rvert}\right)$
`grie`:	Griewangk	$1 + \sum_{i=1}^d x_i^2/4000 - \prod_{i=1}^d \left(\cos(x_i/\sqrt{i})\right)$
`whit`:	Whitley	$-x \sin\left(\sqrt{\lvert x - z \rvert}\right) - z \sin\left(\sqrt{\lvert z + x/2 \rvert}\right)$,
		with $z = y + 47$
`l1`:	L1-Norm	$\sum_i^d \lvert x_i \rvert$
`abs`:	absolute value function	$\lvert x \rvert$
`id`:	identity function	x
`boha`:	Bohachevsky	$x^2 + 2y^2 - 0.3\cos(3\pi x) - 0.4\cos(4\pi y) + 0.7$
`bilcos`:	bisecting line cosine	$x - \cos(\pi x)$

(Rosenbrock) Minimum $x_i^* = (1,1)$. Optimum $f^* = 0$. Starting point $x^{(0)} = (-1.2, 1)$. This is the famous two-dimensional "banana valley" function:

$$100(x_1^2 - x_2)^2 + (1 - x_1)^2.$$

Some authors use a "generalized" Rosenbrock function defined as

$$\sum_{i=1}^{d-1} \left[100(x_{i+1} - x_i^2)^2 + (1 - x_i)^2\right]. \tag{4.2}$$

(Schwefel) Starting point $x^{(0)} = 0$. This function is Schwefel's problem 2.26, a slightly modified variant of Schwefel's problem 2.3 (Schwefel 1995). Both problems are one-dimensional test functions:

$$-x \sin\left(\sqrt{\lvert x \rvert}\right).$$

Besides infinitely many local optima, these functions have a machine-dependent *apparent global optimizer* x_{ap}^*. Schwefel reported that most algorithms located the first or highest local minimum left or right of

$x^{(0)}$. A (10, 100) evolution strategy was able to reach the apparent global optimum x^*_{ap} almost always. Obviously, one dimension is sufficient to demonstrate this effect. However, a d-dimensional variant ($d \geq 1$): $\sum_{i=1}^{d} -x_i \sin\left(\sqrt{|x_i|}\right)$ can be found in the literature, i.e., Whitley et al. (1996).

4.2.3 Test Function for Schwefel's Scenario 3

The L1-norm was used in the third scenario:

(L1-Norm) This function is Schwefel's problem 3.4 (and problem 2.20):

$$\sum_{i=1}^{d} |x_i|.$$

These scenarios will be reconsidered in Chap. 7. Note that the experimenter's skill is needed to set up test functions for optimization scenarios as presented above.

4.3 Elevator Group Control as a Real-World Optimization Problem

Computer simulations are a suitable means to optimize many actual real-world problems. Consider, e.g., a sequence of traffic signals along a certain route or elevators' movements in high-rise buildings. *Optimization via simulation* subsumes all problems in which the performance of the system is determined by running a computer simulation. As the result of a simulation run is a random variable, we cannot optimize the actual value of the simulation output, or a singular performance of the system Y. One goal of optimization via simulation is to optimize the expected performance $E[Y(x_1, x_2, \ldots, x_d)]$, where the x_i's denote the controllable input variables (Schwefel 1979; Azadivar 1999; Banks et al. 2001). The stochastic nature of the simulation output forces the optimization practitioner to apply different methods than are applied in the deterministic counterparts. The stochastic output in optimization via simulation complicates the selection process in direct search methods. The efficiency of the evaluation and selection method is a crucial point, since the search algorithm may not be able to make much progress if the selection procedure requires many function evaluations.

4.3.1 The Elevator Supervisory Group Controller Problem

The construction of elevators for high-rise buildings is a challenging task. Today's urban life cannot be imagined without elevators. The elevator group controller is a central part of an elevator system. It assigns elevator cars to service calls in real time while optimizing the overall service quality, the traffic throughput, and/or the energy consumption. The *elevator supervisory group control* (ESGC) problem can be classified as a combinatorial optimization problem (Barney 1986; So & Chan 1999; Markon et al. 2001). It reveals the same complex behavior as many other stochastic traffic control problems such as materials handling systems with automated guided vehicles. Because of the many difficulties in analysis, design, simulation, and control, the elevator optimization problem has been studied for a long time. First approaches were mainly based on analytical methods derived from queuing theory. Today, *computational intelligence* (CI) methods and other heuristics are accepted as the state of the art (Bäck et al. 1995; Crites & Barto 1998; Schwefel et al. 2003). The elevator group controller determines the floors where the cars should go to. Passengers requesting service can give hall calls. Since the group controller is responsible for the allocation of elevators to hall calls, a control strategy to perform this task in an optimal manner is required. The main goal in designing a better controller is to minimize the time passengers have to wait until they can enter an elevator car after having requested service. This time span is called the *waiting time.*

During a day, different traffic patterns can be observed. For example, in office buildings, an *up-peak traffic* is observed in the morning, when people start to work, and, symmetrically, *down-peak traffic* is observed in the evening. Most of the day there is *balanced traffic* with much lower intensity than at peak times. *Lunchtime traffic* consists of two (often overlapping) phases where people first leave the building for lunch or head for a restaurant floor, and then get back to work (Markon 1995). The ESGC problem subsumes the following task:

> How to assign elevators to passengers in real time while optimizing different elevator configurations with respect to overall service quality, traffic throughput, energy consumption, etc.

Figure 4.2 illustrates the dynamics in an elevator system. Fujitec, one of the world's leading elevator manufacturers, developed a controller that uses a *neural network* (NN) and a set of fuzzy controllers. The weights on the output layer of the neural network can be modified and optimized. The associated optimization problem is quite complex, because it requires the identification of globally optimal NN weights. A further analysis (not shown here) reveals that the distribution of local optima in the ESGC search space is unstructured and that there are many flat plateaus. A *plateau* is a region of candidate solutions with identical function values. For a given candidate solution $x_0 \in \mathbb{R}^d$ exists an ϵ-environment $B(x_0, \epsilon)$ such that $f(x_0) = f(x) \; \forall x \in B(x_0, \epsilon)$.

The objective function values are stochastically disturbed due to the nondeterminism of service calls and are dynamically changing with respect to traffic loads. In general, ESGC research results are not comparable, since the elevator group control per se is not appropriate as a benchmark problem:

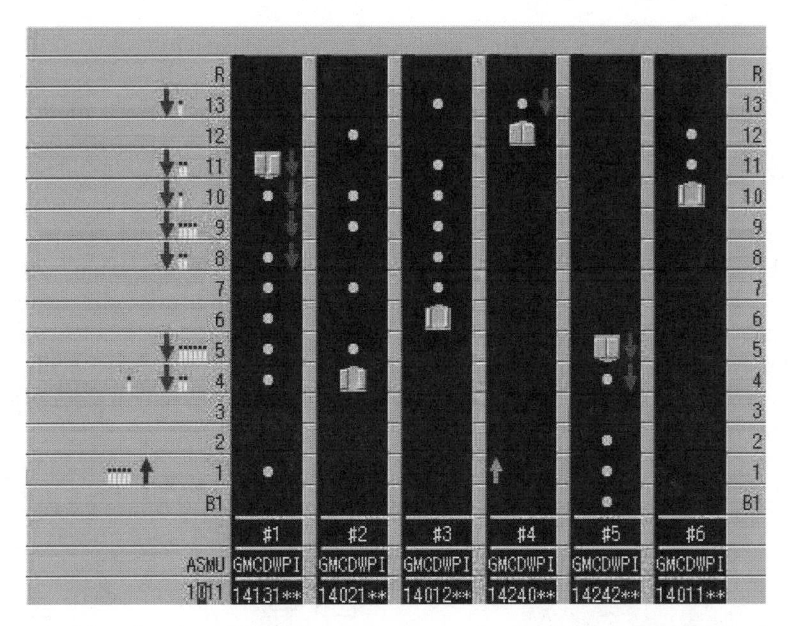

Fig. 4.2. Visualization of the dynamics in an elevator system. Fujitec's elevator simulator representing the fine model. Six elevator cars are serving 15 floors. This model is computationally expensive and has a high accuracy (Beielstein et al. 2003a)

- Elevator systems have a very large number of parameters that differ widely among buildings, elevator models, manufacturers, etc.
- Elevator cars have complex rules of operation, and even slight differences, e.g., in door operation or in the conditions for changing the traveling direction, can affect the system performance significantly. Even small elevator systems have a very large state space, making direct solution infeasible, thus no exact solutions are available for comparison. The sophisticated ESGC rules are usually trade secrets of the manufacturers and cannot be made commonly available for research.

In principle, the optimization practitioner can cope with the enormous complexity of the ESGC problem in two different ways: (i) The problem can be simplified, or (ii) resources can be used extensively. A parallel approach that makes extensive use of a batch-job processing system is presented in Beielstein et al. (2003b). We will concentrate on the first strategy and present a simplified ESGC model. Ideally, a simplified ESGC model should comply with the following requirements: It should enable fast and reproducible simulations and should be applicable to different building and traffic configurations. Furthermore, it must be a valid simplification of a real elevator group controller and thus enable the optimization of one specific controller *policy* π and the comparison of different controller policies. The simplified model should be scalable to enable the simulation of different numbers of floors or servers.

It should be extensible, so that new features (i.e., capacity constraints) can be added. Last but not least, the model is expected to favor experimental and theoretical analyses. In the following section we propose a model—the so-called S-ring—that conforms to all these requirements.

The approach presented next uses two models, an elevator simulator and the S-ring as a simplified model. It is related to *space mapping techniques*. Space mapping techniques iteratively update and optimize surrogate models (Bandler et al. 2004). They use two models for optimization, one fine model (Fujitec's simulator, see Fig. 4.2) that is computationally expensive and has a high accuracy, and one coarse (surrogate) model (the S-ring, see Fig. 4.3) that is fast to solve, but less accurate. The goal is to achieve an improved solution with a minimal number of expensive function evaluations.

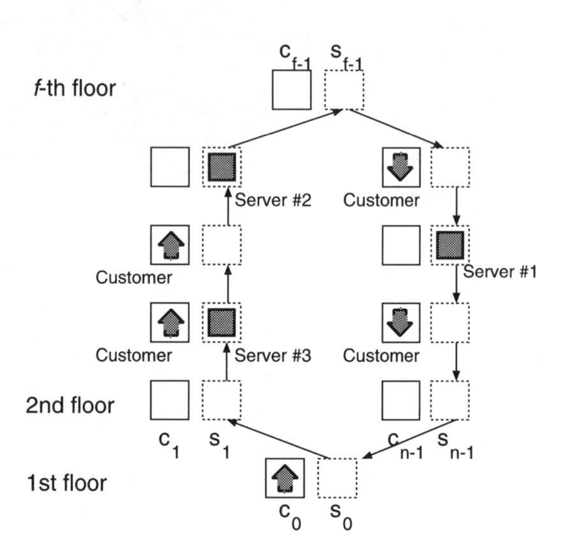

Fig. 4.3. The S-ring as an elevator system. Three cars are serving six floors (or ten sites). The sites are numbered from 0 to $n-1$. There are $f = n/2-1$ floors. This is a coarse (surrogate) model that is fast to solve, but less accurate. Results obtained from this model should be transferable to other systems (Markon et al. 2001)

4.3.2 A Simplified Elevator Group Control Model: The S-Ring

When passengers give a hall call, they simply press a button. Therefore, only a one-bit information for each floor is sent to the ESGC. It appears intuitively correct to map the whole state of the system to a binary string. The system dynamics is represented by a state transition table and can be controlled by a policy. The *sequential-ring model* (S-ring model) has only a few parameters: The number of elevator *cars* (also called servers) m, the number of *sites* n, and the *passenger arrival rate* p (Markon et al. 2001). A 2-bit state (s_i, c_i) is associated with each site. The s_i bit is set to 1 if a server is present on the ith floor, and to 0 otherwise. Correspondingly, the c_i bit is set to 0 or 1 if there is no waiting passenger, respectively at least one waiting passenger. The *state of the system* at time t is given as

$$x(t) := (s_0(t), c_0(t), \ldots, s_{n-1}(t), c_{n-1}(t)) \in \mathbb{B}^{2n}, \tag{4.3}$$

with $\mathbb{B} := \{0, 1\}$.

Example 4.1 (State of the S-ring system). The vector

$$x(t) = (0, 1, 0, 0, \ldots, 0, 1, 0, 0)^T$$

represents the state of the system that is shown in Fig. 4.3. For example, there is a customer waiting on the first floor ($c_0 = 1$), but no server present ($s_0 = 0$). ■

A *state transition table* (Table 4.2) is used to model the dynamics in the system. The state evolution is sequential, scanning the sites from $n-1$ down to 0, and then again around from $n-1$. The up and down elevator movements can be regarded as a loop. This motivates the ring structure. At each time step, one of the sites (floor queues) is considered, where passengers may arrive with probability p.

Table 4.2. The triple $\xi(t) = (c_i, s_i, s_{i+1})$ in the *first column* represents the state of the current site: customer waiting, server present, and server present on the next floor. The probability of a state change to the state $\xi(t+1)$ shown in the *fourth column* is given in the *second column*. *Column three* denotes the decision. The change in the number of sites with waiting customers $\Delta = Q(t+1) - Q(t)$, see Eq. (4.4), is shown in *column five*. That is, the server has to make a decision π (to take or to pass the customer) if there is a customer waiting (1xx), and if there is a server present on the same floor (11x) but no server on the next floor (110). Columns representing configurations in which the policy affects the state of the systems are shaded *dark gray*; the configuration of the take decision from Example 4.2 is printed in *boldface*

$\xi(t)$	Prob	$\pi(x)$	$\xi(t+1)$	Δ
000	$1-p$	$\{0,1\}$	000	0
000	p	$\{0,1\}$	100	+1
001	$1-p$	$\{0,1\}$	001	0
001	p	$\{0,1\}$	101	+1
010	$1-p$	$\{0,1\}$	001	0
010	p	0	101	+1
010	p	1	010	0
011	1	$\{0,1\}$	011	0
100	1	$\{0,1\}$	100	0
101	1	$\{0,1\}$	101	0
110	1	0	101	0
110	**1**	**1**	**010**	**−1**
111	1	$\{0,1\}$	011	−1

Example 4.2 (S-ring). Consider the situation at the third site (the up direction on the third floor) in Fig. 4.3. As a customer is waiting, a server is present,

and there is no server on the next floor, the controller has to make a decision. The car can serve the customer (take decision), or it can ignore the customer (pass decision). The former would change the values of the corresponding bits from $(1, 1)$ to $(1, 0)$, and the latter from $(1, 1)$ to $(0, 1)$. ■

As the rules of operation are very simple, this model is easily reproducible and is suitable for benchmark testing. Despite the model's simplicity, it is hard to find the optimal policy π^* even for a small S-ring. The real π^* is not obvious, and its difference from heuristic suboptimal policies is nontrivial.

So far, the S-ring has been described as a simulation model. To use it as an optimization problem, it is equipped with an objective function. Consider the function that counts the sites with waiting customers at time t:

$$Q(t) = \hat{Q}(x, t) = \sum_{i=0}^{n-1} c_i(t). \tag{4.4}$$

Then the steady-state time-average number of sites with waiting customers in the queue is

$$\mathcal{Q} = \lim_{T \to \infty} \frac{\int_0^T Q(t) \mathrm{d}t}{T}, \quad \text{with probability one.} \tag{4.5}$$

The basic optimal control problem is to find a policy π^* for a given S-ring configuration. The optimal policy minimizes the expected number of sites with waiting passengers in the system, that is, the steady-state time-average as defined in Eq. (4.5). A $2n$-dimensional vector, $y \in \mathbb{R}^{2n}$, can be used to represent the policy. Let $\theta : \mathbb{R} \to \mathbb{B}$ define the Heaviside function:

$$\theta(z) = \begin{cases} 0, \text{ if } z < 0, \\ 1, \text{ if } z \geq 0, \end{cases} \tag{4.6}$$

and $x = x(t)$ be the state at time t (see Eq.(4.3)). A linear discriminator, or perceptron,

$$\pi(x) = \pi(x, y) = \theta \langle y, x \rangle, \tag{4.7}$$

can be used to present the policy in a compact manner. For a given vector y that represents the policy, and a given vector x that represents the state of the system, a *take* decision occurs if $\pi(x, y) \geqslant 0$, otherwise the elevator will ignore the customer.

The most obvious heuristic policy is the *greedy* one: When given the choice, always serve the customer. The $2n$-dimensional vector $y_{\text{greedy}} = (1, 1, \dots, 1)^T$ can be used to represent the greedy policy. This vector guarantees that the product in Eq. (4.7) equals 1, which is interpreted as a take decision. Rather counterintuitively, this policy is not optimal, except in the heavy traffic ($p > 0.5$) case. This means that a good policy must bypass some customers occasionally to prevent a phenomenon that is known as *bunching*, which occurs in elevator systems when nearly all elevator cars are positioned in close proximity to each other.

The perceptron S-ring problem can serve as a benchmark problem for many optimization algorithms, since it relies on the fitness function:

$$F : \mathbb{R}^{2n} \to \mathbb{R}$$

(Markon et al. 2001; Beielstein & Markon 2002). Figure 4.4 shows the correlation between the noisy function values and the estimated function values. Bartz-Beielstein et al. (2005c) describe the S-ring model as a partially observable Markov decision process in detail.

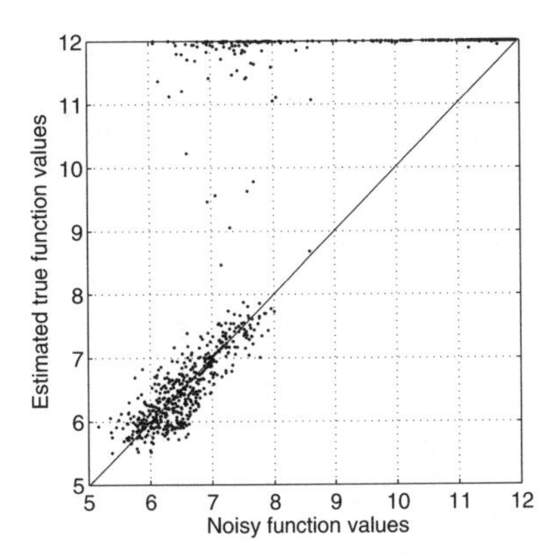

Fig. 4.4. S-ring. Estimated versus noisy function values. Test instance `sring24` as listed in Table 4.3. Estimated values have been gained through reevaluation, whereas noisy function values are based on one evaluation only. Points representing values from functions without noise would lie on the bisector

4.3.3 The S-Ring Model as a Test Generator

The S-ring model can be used to generate test problem instances. An S-ring problem instance can be characterized by the number of sites, the number of elevator cars, the arrival probability, and the simulation time. The S-ring model has been used by Markon et al. (2006) to generate test instances as shown in Table 4.3. A problem design specifies one or more instances of an optimization problem and related restrictions, i.e., the number of available resources (function evaluations). In addition, a computer experiment requires

Table 4.3. Test instances for the S-ring model

Instance	Dimension	Number of sites	Number of elevator cars	Arrival probability	Simulation time
`sring12`	12	6	2	0.2	1000
`sring24`	24	12	4	0.2	1000
`sring36`	36	18	8	0.2	1000
`sring48`	48	24	16	0.3	1000
`sring96`	96	48	32	0.3	1000

the specification of an algorithm design. As designs play an important role in experimentation, they will be discussed in the following chapter.

4.4 Randomly Generated Test Problems

Although the S-ring model can be used to generate problem instances at random, these instances have been generated deterministically. Three important problems related to randomly generated problem instances can be mentioned:

Problem 4.1 (Floor or ceiling effects). Rardin & Uzsoy (2001) illustrate subtle and insidious pitfalls that can arise from the randomness of the instance generation procedure with a simple example: To generate instances of the n-point traveling salesperson problem (TSP), $(n \times n)$ symmetric matrices of point-to-point distances are generated as follows:

> Fill the upper triangle of an $n \times n$ cost matrix with $c_{i,j}$ generated randomly (independently and uniformly) between 0 and 20. Then complete the instance by making $c_{j,i} = c_{i,j}$ in the lower triangle and setting $c_{i,i} = 0$ along the diagonal.

The mean of the cell entries $c_{i,j}$ with $i < j$ is 10 with standard deviation 5.77. If $n = 5000$ points are generated, the average tour length will be $10 \cdot 5000 = 50,000$ with standard deviation $5.77\sqrt{5000} = 408$. Nearly every feasible tour will have a length within $\pm 3 \cdot 408$ of $50,000$. Hence, "almost any random guess will yield a good solution."

Problem 4.2 (Search-space structure). Reeves & Yamada (1998) report that local optima of randomly generated permutation flow-shop scheduling problem instances are distributed in a *big-valley structure*, i.e., local optima are relatively close to other local optima. This big-valley structure in the search space topology is well-suited for many optimization algorithms. But do structured problem instances, which are assumed to be more realistic, possess a similar distribution? Watson et al. (1999) showed for permutation flow-shop scheduling problems that local optima are generally distributed on large plateaus of equally fit solutions. Therefore, the assumption of big-valley structured local optima distributions does not hold for this type of problem. Whitley et al. (2002) conclude that there are differences in the performance of scheduling algorithms on random and structured instances.

The distribution of the S-ring local optima is not purely random. An analysis of the search space shows that there are plateaus of equally good solutions.

Problem 4.3 (Different sources of randomness). To separate different sources of randomness is a basic principle in statistics. Equation (3.9) describes how the total variability can be partitioned into its components:

$$SS_{\mathrm{T}} = SS_{\mathrm{TREAT}} + SS_{\mathrm{E}}.$$

This will be referred to as the *fundamental* ANOVA *principle* in the following.

If stochastic search algorithms are subject of the analysis, using randomly generated test instances will add another source of randomness to the algorithm's randomness that might complicate the analysis.

4.5 Recommendations

Standard test suites are valuable tools in the first phase of an experimental analysis.

Clearly specified hypotheses should be used in the second phase of the analysis, e.g., based on the guidelines from experimental algorithmics (GL 2.1 to 2.5) and Mayo's extensions as presented in Sect. 2.5. Statistical tests are used as learning tools, they provide means to evaluate what has been learned.

4.6 Summary

The basic ideas from this chapter can be summarized as follows:

1. Specifying test functions, performing tests, measuring performances, and selecting the algorithm with the best performance is a commonly used procedure.
2. Not only the set of test functions, but also the set of test instances has to be chosen carefully.
3. Test functions can be distinguished from real-world optimization problems.
4. Test functions should be combined with an optimization scenario.
5. The S-ring model defines a simplified elevator group control problem. It
   (a) enables fast and reproducible simulations,
   (b) is applicable to different buildings and traffic patterns,
   (c) is scalable and extensible, and
   (d) can be used as a test problem generator.
6. A problem design specifies at least one problem instance plus related restrictions.
7. Randomly generated problem instances can complicate the analysis of stochastic search algorithms.

4.7 Further Reading

Whitley et al. (1996) provide a good starting point for the discussion of test functions in evolutionary computation. Schwefel (1995) presents a thoughtfully compiled set of test functions to compare deterministic and stochastic optimization algorithms. Elevator group control and related problems are analyzed in Markon et al. (2006).

5

Designs for Computer Experiments

> A common mistake people make when trying to design something completely foolproof is to underestimate the ingenuity of complete fools.
>
> —Douglas Adams

This chapter discusses designs for computer experiments. Before the optimization runs are started, the experimenter has to choose the parameterizations of the optimization algorithm and one or more problem instances.

Johnson (2002) suggests to explain the corresponding adjustment process in detail, and therefore to include the time for the adjustment in all reported running times to avoid a serious underestimate. An important step to make this procedure more transparent and more objective is to use design of experiments techniques. They provide an algorithmic procedure to tune the exogenous parameter settings for the algorithms under consideration before the complex real-world problem is optimized or two algorithms are compared. Experimental design provides an excellent way of deciding which simulation runs should be performed so that the desired information can be obtained with the least number of experiments (Box et al. 1978; Box & Draper 1987; Kleijnen 1987; Kleijnen & Van Groenendaal 1992; Law & Kelton 2000).

We will develop experimental design techniques that are well suited for parameterizable search algorithms such as evolution strategies, particle swarm optimization, or Nelder–Mead simplex algorithms. The concept of splitting experimental designs into algorithm and problem designs, which was introduced for evolution strategies in Beielstein et al. (2001), is detailed in the following. Algorithm tuning as introduced in Chap. 7 refers to the task of finding an optimal (or improved) algorithm design for one specific problem design.

Design decisions can be based on geometric or on statistical criteria. Regarding geometric criteria, two different design techniques can be distinguished: The samples can be placed either (1) on the boundaries, or (2) in the interior of the design space. The former technique is used in DOE, whereas DACE uses the latter approach. An experiment is called *sequential* if the experimental conduct at any stage depends on the results obtained so far. Sequential approaches exist for both variants. We recommend using factorial designs or space-filling designs instead of the commonly used one-factor-at-a-time designs. It is still an open question which design characteristics are important: "... extensive empirical studies would be useful for better understanding what sorts of designs perform well and for which models"(Santner et al. 2003, p. 161).

5.1 Computer Experiments

Optimization runs will be treated as experiments. There are many degrees of freedom when starting an optimization run. In many cases search algorithms require the determination of parameters such as the population size in evolutionary algorithms before the optimization run is performed. From the viewpoint of an experimenter, design variables (factors) are the parameters that can be changed during an experiment. Generally, there are two different types of factors that influence the behavior of an optimization algorithm:

1. problem-specific factors, i.e., the objective function
2. algorithm-specific factors, i.e., the population size or other exogenous parameters

We will consider experimental designs that comprise problem-specific factors and exogenous algorithm-specific factors. Algorithm-specific factors will be considered first. *Endogenous* can be distinguished from *exogenous parameters* (Beyer & Schwefel 2002). The former are kept constant during the optimization run, whereas the latter, e.g., standard deviations, are modified by the algorithms during the run. Standard deviations will be referred to as *step widths* or *mutation strengths*. Considering particle swarm optimization, step widths and their associated directions are frequently referred to as *velocities*.

An *algorithm design* $X_A \subseteq \mathcal{D}_A$ ($\mathcal{D}_A$ denotes the space of all algorithm designs, i.e., all possible exogenous parameter settings) is a set of vectors, each representing one specific setting of the design variables of an algorithm. A design can be specified by defining ranges of values for the design variables. Note that a design can contain none, one, several, or even infinitely many design points.

Example 5.1 (Algorithm design). Consider the set of exogenous strategy parameters for particle swarm optimization algorithms with the following values: swarm size $s = 10$, cognitive parameter $c_1 \in [1.5, 2]$, social parameter $c_2 = 2$, starting value of the inertia weight $w_{\max} = 0.9$, final value of the inertia weight $w_{\text{scale}} = 0$, percentage of iterations for which $w_{\max}$ is reduced $w_{\text{iterScale}} = 1$, and maximum value of the step size $v_{\max} = 100$. This algorithm design contains infinitely many design points. ■

The *optimal algorithm design* is denoted as X_A^*. Optimization is interpreted in a very broad sense—it can refer to the best design point x_a^* as well as the most informative design points. *Problem designs* $X_P \subseteq \mathcal{D}_P$ ($\mathcal{D}_P$ denotes the space of all instances of one optimization problem) provide information related to the optimization problem, such as the available resources (number of function evaluations) or the problem's dimension.

An *experimental design* $X_E \subseteq \mathcal{D}$ ($\mathcal{D}$ denotes the space of all experimental settings) consists of a problem design X_P and an algorithm design X_A. The run of a stochastic search algorithm can be treated as an experiment with a stochastic output $Y(x_a, x_p)$, with $x_a \in \mathcal{D}_A$ and $x_p \in \mathcal{D}_P$. If the random seed is specified, the output would be deterministic. This case will not be considered further, because it is not a common practice to specify the seed that is used in an optimization run. Performance can be measured in many ways, for example, as the best or the average function value for n runs. One of our goals is to find a design point $x_a^* \in \mathcal{D}_A$ that improves the performance of an optimization algorithm for one problem design point $x_p \in \mathcal{D}_P$. To test the robustness of an algorithm, more than one design point can be considered.

Example 5.2 (Problem design). Robustness can be defined as good performance over a wide range of problem instances. A very simple example is the function `sphere`: $\sum_{i=1}^{d} x_i^2$ and a set of d-dimensional starting points

$$x_i^{(0)} = \left(-i, i, \ldots, (-i)^d\right)^T, \qquad i = 1, 2, 3.$$

■

The optimization of real-world problems requires algorithms with good initial parameters, since many real-world problems are computationally expensive, e.g., *optimization via simulation* (Schwefel 1979; Banks et al. 2001). Therefore only a few optimization runs are possible, which should be performed with good parameter settings. Optimization practitioners are interested in obtaining a good parameter setting with a minimum number of optimization runs. The choice of an adequate parameter setting, or design, can be based on expert knowledge. But in many cases there is no such knowledge available.

5.2 Classical Algorithm Designs

In this section we will consider the following task: Determine an improved algorithm design point $x_a^* \in \mathcal{D}_A$ for one fixed problem design point $x_p \in \mathcal{D}_P$.

Consider the regression model $y = X\beta + \epsilon$ that was defined in Eq. (3.12) with associated regression matrix X as introduced in Eq. (3.13). The regression matrix X is referred to as the *design matrix* in the context of experimental designs. The optimal design can be understood as the set of input vectors $X^* \subset \mathcal{D}_A$ that generates output values y that are as informative as possible with respect to the exact functional relationship (Eq. (3.11)). Hence, the optimal algorithm design provides more information than any other algorithm design with respect to some optimality criterion. This information can be used to detect an improved design point.

The classical criteria for optimality such as D-optimality have to cope with the dependence on the model parameters. These so-called *alphabetic optimal designs* attempt to choose design points so that some measure of error in prediction, which depends on the underlying assumed model, is minimized (Federov 1972; Box & Draper 1987; Pukelsheim 1993; Spall 2003).

Example 5.3 (Optimality criteria).

1. A design is A-optimal if it minimizes the sum of the main diagonal elements of $(X^TX)^{-1}$. Hence, as can be seen from Eq. (3.15), A-optimal designs minimize the sum of the variances of the regression coefficients.
2. A design is said to be D-optimal if

$$\det\left((X^TX)^{-1}\right) \tag{5.1}$$

is minimized, where X is the design matrix (Montgomery 2001). ■

Often, it is not trivial to formulate the experimental goals in terms of these optimal design criteria. And, "even if we can formulate the problem in this way, finding the optimal design may be quite difficult"(Santner et al. 2003, p. 124). Despite of these problems, factorial designs as one relevant and often applied type of D-optimal designs will be introduced in the following section.

Factorial Designs

The commonly used *one-factor-at-a-time* method, where certain factors are varied one at a time, while the remaining factors are held constant, provides an estimate of the influence of a single parameter at selected fixed conditions of the other parameters. Such an estimate may only have relevance under the assumption that the effect would be the same at other settings of the other parameters. This requires that effects of variables behave additively on the response over the ranges of current interest. Furthermore, interactions cannot be determined. Therefore, we do not recommend using this method.

Factorial designs are more efficient than one-factor-at-a-time designs (Kleijnen 1987). Box et al. (1978) give an instructive example that explains the weakness of the classical one-factor-at-a-time design. Orthogonal designs simplify the computations. They lead to uncorrelated regression coefficients ($cov(\beta_i, \beta_j) = 0$, cf. Eq. (3.15)) and to a minimal variance of the predicted response in the design space.

In the following, we use orthogonal designs with two levels for each factor: The corresponding factorial design with k factors requires 2^k experimental runs. This is a 2^k *full factorial design* or simply a 2^k design. Since interactions that involve many factors can be neglected in some situations, *fractional factorial designs* omit the corresponding run configurations and require only 2^{k-p} runs. Adding center points and axial points to 2^k designs leads to *central composite designs* (CCD) with axial runs (Fig. 5.1). The values of factor

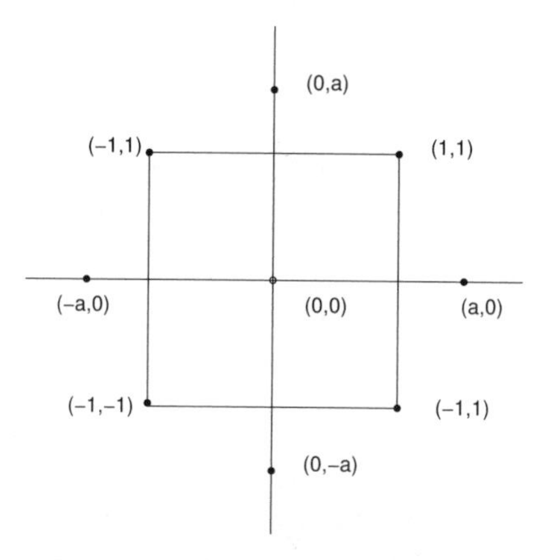

Fig. 5.1. Central composite design with axial runs for $k = 2$. The value of $a = \sqrt{k}$ gives a spherical CCD; that is, all factorial and axial design points are on the surface of a sphere of radius $\sqrt{k}$ (Montgomery 2001)

levels can be scaled. A variable x is called *scaled* or standardized if x ranges between -1 and $+1$.

An important objection against 2^k designs is that nonlinear effects remain undiscovered. Therefore, 2^k designs are only used to get an overview over the effects and their interactions, not to obtain the exact values. Furthermore, techniques to measure the goodness of the model fit can be applied (Montgomery 2001).

Hence, the entry -1 in the regression matrix X (Eq. (3.13)) denotes a factor at its low level, and $+1$ a factor at its high level. Table 5.1 depicts a fractional factorial 2^{9-5}_{III} design.

In general, the following two purposes require different designs:

1. Factorial designs are used to determine which factors have a significant effect in the screening phase of the DOE.
2. To fine-tune the algorithm in the modeling and optimization phase, CCDs, which extend the factorial designs, can be used.

The number of samples in the CCD scales as 2^k, where k is the number of factors in the model. Therefore CCD should only be used in the final phase of the DOE procedure when the number of factors is very low.

Factorial designs that are commonly used in classical DOE place samples on the boundaries of the design space. The interior remains unexplored. This is due to the following model assumptions: The underlying model in the classical DOE approach can be written as

$$\tilde{y} = y + \epsilon, \tag{5.2}$$

where $\tilde{y}$ is the measured response, y the true value, and ϵ an error term. The errors are usually assumed to be independent and identically distributed.

Table 5.1. Fractional factorial 2^{9-5}_{III} design. This design is used for screening the ES parameters. Concrete values are shown in Table 7.4

Conf.	A	B	C	D	E=ABC	F=BCD	G=ACD	H=ABD	J=ABCD
1	−	−	−	−	−	−	−	−	+
2	+	−	−	−	+	−	+	+	−
3	−	+	−	−	+	+	−	+	−
4	+	+	−	−	−	+	+	−	+
5	−	−	+	−	+	+	+	−	−
6	+	−	+	−	−	+	−	+	+
7	−	+	+	−	−	−	+	+	+
8	+	+	+	−	+	−	−	−	−
9	−	−	−	+	−	+	+	+	−
10	+	−	−	+	+	+	−	−	+
11	−	+	−	+	+	−	+	−	+
12	+	+	−	+	−	−	−	+	−
13	−	−	+	+	+	−	−	+	+
14	+	−	+	+	−	−	+	−	−
15	−	+	+	+	−	+	−	−	−
16	+	+	+	+	+	+	+	+	+

Eq. (5.2) is used to model the assumption that ϵ is always present. Therefore the goal of classical DOE is to place samples in the design space so as to minimize its influence. DOE employs an approximation model,

$$\hat{y} = f(x, \tilde{y}(x)), \tag{5.3}$$

where f is usually a low-order polynomial, and x denotes a sample point. We can conclude from these model assumptions that design points should be placed on the boundaries of the design space. This can be seen in Fig. 5.2: The random errors remain the same in both design configurations, but the estimated linear model (dotted lines) gives a poor approximation of the true model if the samples are located in the interior of the design space (left figure). Moving the design points to the boundaries, as shown in the right figure, yields a better approximation of the true relationship.

5.3 Modern Algorithm Designs

Replicate runs reduce the variance in the sample means and allow the estimation of the random error ϵ in stochastic computer experiments, cf. Eq. (5.2).

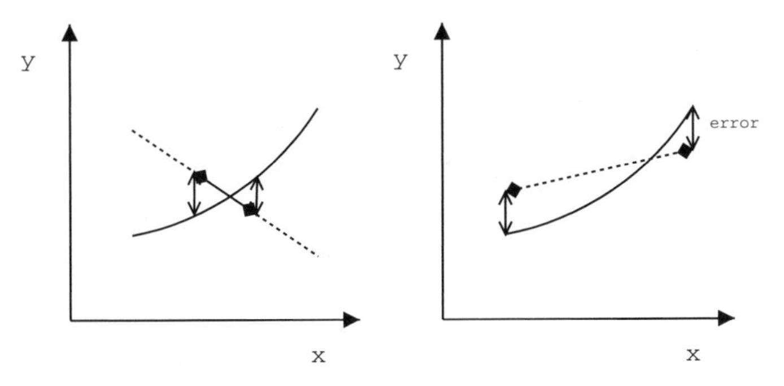

Fig. 5.2. DOE approximation error. The errors and true models (*solid lines*) are the same in both configurations. Moving the design points to the boundaries yields a better approximation model (*dotted lines*) (Trosset & Padula 2000)

Modern DACE methods have been developed for deterministic computer experiments that have no random error. DACE assumes that the interesting features of the true model can be found in the whole sample space. Therefore, space-filling or exploratory designs, which place a set of samples in the interior of the design space, are commonly used.

Metropolis & Ulam (1949) introduced a pseudo-Monte Carlo sampling method for computer simulations. As *Monte Carlo sampling* (MC) places samples randomly in the design space, large regions may remain unexplored. Stratified MC sampling divides the design space into subintervals of equal probabilities and therefore requires at least 2^d samples.

McKay et al. (1979) proposed *Latin hypercube sampling* (LHS) as an alternative to Monte Carlo sampling. The resulting designs are called *Latin hypercube designs* (LHD). LHS is superior under certain assumptions to MC sampling and provides a greater flexibility in choosing the number of samples. LHS can be used to generate points for algorithm designs. One instance of a LHD with ten design points in two dimensions is shown in Fig. 5.3. Note that LHS might result in an ill-designed arrangement of sample points, for example, if the samples are placed along a diagonal as shown in Fig. 5.3.

Example 5.4 (Ill-designed arrangements). Santner et al. (2003) describe the following consequences that arise from ill-designed arrangements. Consider the function

$$y(x_1, x_2) = \frac{x_1}{1 + x_2}, \quad (x_1, x_2) \in [0, 1] \times [0, 1]. \tag{5.4}$$

The stochastic process was chosen as $Y = \beta_0 + Z$, where $Z(\cdot)$ is a Gaussian stochastic process with zero mean, unknown process variance, and power exponential correlation function (Eq. (3.18)). A comparison of the prediction errors of the two designs from Fig. 5.3 yields "a better predictor over most of the design space except for the diagonal," where the ill-arranged design "collects most of its data." ■

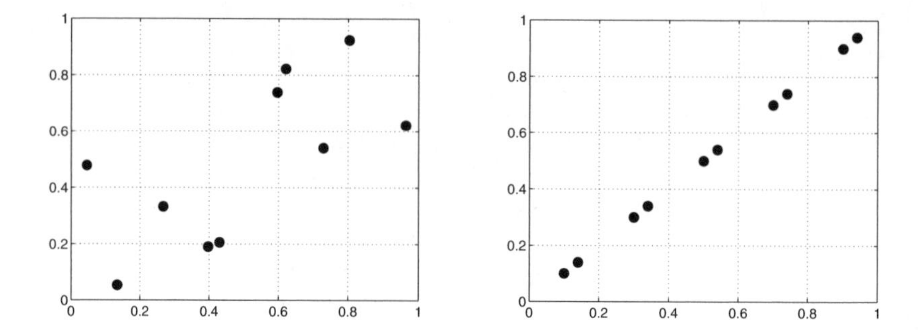

Fig. 5.3. Two LHD samples of ten points. *Left:* Typical instance of a LHD. *Right:* This design is considered an ill-designed arrangement, because it is not "truly space-filling" (Santner et al. 2003, p. 130)

In addition, Santner et al. (2003) discuss several criteria that can be applied to generate designs by distance-based criteria, for example, maxmin distance designs, or measures of the discrepancy between the empirical distribution of the set of sample points and the uniform distribution.

5.4 Sequential Algorithm Designs

In some situations, it might be beneficial to generate the design points not at once, but *sequentially*. The selection process for further design points can include knowledge from the evaluation of previously generated design points.

Evolutionary operation (EVOP) was introduced in the 1950s by Box (1957). The basic idea is to replace the static operation of a process by a systematic scheme of modifications (*mutations*) in the control variables. The effect of these modifications is evaluated and the process is shifted in the direction of improvement: The best *survives*. Box associated this purely deterministic process with an organic mutation-selection process. Satterthwaite (1959a, b) introduced a *random evolutionary operation* (REVOP) procedure. REVOP was rejected by Box because of its randomness.

Sequential designs can be based on several criteria, for example, on the D-optimal maximization criterion as presented in Example 5.3. We will present a sequential approach that is based on a criterion developed for DACE, the *expected improvement*.

Sequential sampling approaches with adaptation have been proposed for DACE metamodels. For example, Sacks et al. (1989) classified sequential sampling approaches with and without adaptation to the existing metamodel. Jin et al. (2002) propose two sequential sampling approaches with adaptation that are not limited to DACE models.

Santner et al. (2003, p. 178) present a heuristic algorithm for unconstrained problems of global minimization. Let $y^n_{\min}$ denote the smallest known minimum value after n runs of the algorithm. The *improvement* is defined as

$$\text{improvement at } x = \begin{cases} y^n_{\min} - y(x), & y^n_{\min} - y(x) > 0, \\ 0, & y^n_{\min} - y(x) \leq 0, \end{cases} \tag{5.5}$$

for $x \in \mathcal{D}_A$. As $y(x)$ is the realization of a random variable, its exact value is unknown. The goal is to optimize its expected value, the so-called *expected improvement* (EXPIMP). The discussion in Santner et al. (2003, p. 178 ff.) leads to the conclusion that new design points are attractive "if either there is a high probability that their predicted output is below the current observed minimum and/or there is a large uncertainty in the predicted output." This result is in accordance with the experimenters' intention to avoid sites that guarantee worse results and to improve the model at the same time. It constituted the motivation for the EXPIMP heuristic shown in Fig. 5.4 (Bartz-Beielstein & Markon 2004; Bartz-Beielstein et al. 2004b). Next, we will discuss problem designs that consider more than just one design point of one problem design.

Heuristic: EXPIMP

1. Choose an initial design $X_A^{(n)} \in \mathcal{D}_A$ with n points.
2. Run the algorithm at $x_i \in X_A^{(n)}$, $i = 1, \ldots, n$, to obtain the vector of output values $y(x)$.
3. Check the termination criterion.
4. Select a new point x_{n+1} that maximizes the expected improvement, cf. Eq. (5.5).
5. Run the algorithm at x_{n+1} to obtain the output $y(x_{n+1})$.
6. Set $X_A^{(n+1)} = X_A^{(n)} \cup \{x_{n+1}\}$, $n = n + 1$, and go to 3.

Fig. 5.4. Expected improvement heuristic

5.5 Problem Designs

Different instances of one optimization problem can be used to compare algorithms. The problem design and the algorithm design to be compared can be arranged in matrix form (Rardin & Uzsoy 2001). This matrix form will be used to present performance measures that consider more than one problem design point simultaneously. These performance measures will be introduced in Sect. 7.2.3.

5.5.1 Initialization

It can be observed that the performance of optimization algorithms depends crucially on the *starting point* $x^{(0)}$. There are mainly two different initialization methods: deterministic and random starts. To test the robustness of algorithms and not only their efficiency, Hillstrom proposed to use a series of random starts. Twenty random starts are considered as "a compromise between sample size and testing expense" (Hillstrom 1977). This initialization method is nowadays often used for stochastic search heuristics such as particle swarm optimization algorithms.

More et al. (1981) state that the use of random starts affects the reproducibility of the results. Furthermore, random starting points introduce an additional source of randomness. Since some methods of our analysis try to explain as much randomness as possible by the differences between the algorithms, this initialization method may cause unwanted side-effects that complicate the statistical analysis. Better suited for our needs are deterministic routines. We will present initialization and termination methods next.

To initialize the set of search points $X^{(0)} = \{x_1^{(0)}, \dots, x_p^{(0)}\}$, the following methods can be used:

(DETEQ) *Deterministic.* Each search point uses the same vector, which is selected deterministically, i.e., $x_{\text{init}} = \mathbf{1}^T \in \mathbb{R}^d$. As this method uses only one starting point x_{init}, it is not suitable to visualize the starting points for which the algorithm converged to the optimum.

Example 5.5. Schwefel (1995) proposed the following initialization scheme for high-dimensional nonquadratic problems:

$$x_i^{(0)} = x^* + \frac{(-1)^i}{\sqrt{d}}, \qquad \text{for } i = 1, \dots, d. \tag{5.6}$$

■

(DETMOD) *Deterministically modified starting vectors.* The algorithm can be tested with starting vectors $x^{(0)}$, $10x^{(0)}$, and $100x^{(0)}$ (More et al. 1981), or any other scheme that generates starting points deterministically.

(UNIRND) *Uniform random starts.* Every search point $(i = 1, \dots, p)$ uses the same vector $x_{\text{init}} \in \mathbb{R}^d$, where the d components are realizations of independent $U[x_l, x_u]$ random variables. This method introduces an additional source of randomness. It is suitable to visualize the starting points for which the algorithm converged to the optimum. This visualization technique is useful to get some insight into the behavior of the algorithm.

(NUNIRND) *Nonuniform random starts.* Every search point uses a different vector $x_i^{(0)}$, $(i = 1, \dots, p)$, that is, $X^{(0)} = \{x_1^{(0)}, \dots, x_p^{(0)}\}$, with $x_i^{(0)} \neq x_j^{(0)} \; \forall \, i \neq j$. Each of the p vectors $x_{\text{init}} \in \mathbb{R}^d$ consists of d components that are realizations of independent $U[x_l, x_u]$ random variables. This initialization method is used by many authors. It introduces an additional source of randomness, and it is not suitable to visualize the starting points for which the algorithm converged to the optimum.

Since variance reducing techniques are considered in our analysis, and we are trying to explain the variance in the results based on the fundamental ANOVA principle (Eq. (3.10)), we prefer the deterministic initialization scheme DETEQ to gain insight into the algorithm's behavior. To test its robustness, randomly or deterministically varied initialization schemes are advantageous.

Example 5.6 (Influence of different initialization schemes). A (1+1)-ES is started for the same problem and algorithm design, only the initialization schemes were modified. The algorithms use the following initialization schemes:

- UNIRND, $x^{(0)} \sim U[a,b]$. Note that for the (1+1)-ES there is no difference between UNIRND and NUNIRND.
- DETEQ, $x^{(0)} = (a+b)/2$.
- DETMOD. The interval $[a,b]$ is divided into $t_{\max}$ equidistant subintervals $I_k = [a_k, b_k]$, $(k = 1, \dots, t_{\max})$. The kth run uses $x^{(0)} = a_k$.

Results from this comparison are shown in Fig. 5.5. ■

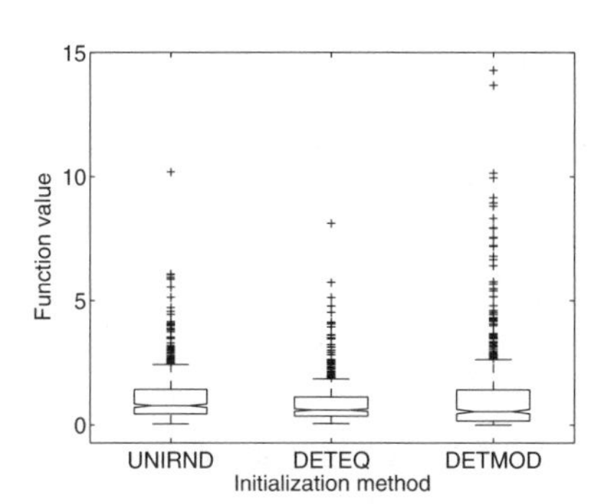

Fig. 5.5. Boxplots illustrating the effect of three different initialization methods ($a = 10$, $b = 100$). The deterministic initialization scheme DETEQ, which uses the same starting point for every run, generates the lowest variance in the data. The DETMOD initialization can be used to test the robustness of the algorithm because it generates the highest variance

5.5.2 Termination

An algorithm run terminates if it (or its budget) is:

(XSOL/FSOL) *Solved.* The problem was solved.

1. A domain convergence test becomes true when the x_i's are close enough in some sense.
2. A function value convergence test becomes true when the function value is close enough in some sense.

(STAL) *Stalled.* The algorithm has stalled. A step-size test becomes true when the step sizes are sufficiently small.

(EXH) *Exhausted.* The resources are exhausted.

1. An iteration test becomes true if the maximum number of function values is exhausted.
2. A no-convergence-in-time test becomes true. This includes domain convergence and function value convergence.

Tests specified for the cases in which the algorithm is stalled or its budget is exhausted are called *fail tests.* Termination is as important as initialization. Even if the algorithm converges in theory, rounding errors may prevent convergence in practice. Thus, fail tests are necessary for every algorithm. Singer & Singer (2004) demonstrate the impact of the termination tests on the performance of a Nelder–Mead or simplex algorithm: "A fairly simple efficiency analysis of each iteration step reveals a potential computational bottleneck in the domain convergence test."

5.6 Discussion: Designs for Computer Experiments

5.6.1 Problems Related to Classical Designs

The assumption of a linear model for the analysis of computer algorithms is highly speculative. As can be seen from Fig. 5.2, besides the selection of a correct regression model, the choice of design points is crucial for the whole procedure. Beielstein (2003) used a linear regression model to determine improved algorithm designs for evolution strategies and simulated annealing. This modeling approach holds more pitfalls than the DACE approach, which uses space-filling designs and a stochastic process model. It will be presented in Chap. 7.

5.6.2 Problems Related to Modern Designs

On the other hand, DACE was introduced for deterministic computer experiments and not for the analysis of stochastic search algorithms. Performing repeated runs and taking the mean value at the design points enables the application of these techniques even for nondeterministic experiments. Determinism is "introduced through the back door."

Another problem that arises from DACE designs is the treatment of qualitative factors. Moreover, as Santner et al. (2003, p. 149) note:

> It has not been demonstrated that LHDs are superior to any designs other than simple random sampling (and they are only superior to simple random sampling in some cases).

5.7 Recommendations

Based on our experience, we can give the following recommendations: If only a few qualitative factors are relevant, then for each setting a separate Latin hypercube design could be used. Otherwise, factorial design could be used to screen out those qualitative factors that have the largest effect during the first step of the experimentation. Latin hypercube designs can be used in the second step to refine the analysis.

Factorial designs are applicable to test specific design points and combinations of specific settings, e.g., a population size $s = 10$ and a mutation strength $\sigma \in \{1, 2\}$, whereas space-filling designs are well suited to test parameter ranges, e.g., $s \in [10, 20]$ and a $\sigma \in [1, 2]$. In general, we prefer space-filling designs to model the effects and interactions of factors from optimization algorithms.

Despite the recommendations given in this chapter, the most frequently used strategy in practice will be the *best-guess strategy*. It works reasonably well in many situations, because it benefits from the experimenter's feeling or skill.

> In England it is still not uncommon to find in a lab a youngish technician, with no formal education past 16 or 17, who is not only extraordinarily skillful with the apparatus, but also the quickest at noting an oddity on for example the photographic plates he has prepared from the electron microscope (Hacking 1983).

Relying upon high-level experimental design theories may sometimes "help" the experimenter to miss the point.

5.8 Summary

The results from this chapter can be summarized as follows:

1. An experimental design consists of a problem design and an algorithm design.
2. Algorithm designs consider only exogenous strategy parameters.
3. Endogenous strategy parameters are modified during the run of an algorithm.
4. The objective function, its dimension, and related constraints are specified in the problem design $X_P \subseteq \mathcal{D}_P$.
5. The algorithm design $X_A \subseteq \mathcal{D}_A$ defines the set of exogenous strategy parameters of the algorithm, for example, the swarm (population) size of a particle swarm optimization.
6. The task of searching for an optimized algorithm design for a given problem design is called algorithm tuning.
7. We do not recommend using one-factor-at-a-time designs, because they fail to discover any possible interaction between the factors.
8. Factorial designs are widely used designs from classical DOE. Design points are placed on the boundaries of the design space.

9. Latin hypercube designs are popular designs for modern DACE. These designs are space filling: Design points are placed in the interior of the design space.
10. Sequential designs can be constructed for both classical and modern designs.
11. Designs of test problems specify one specific problem instance. This specification comprises the starting conditions and the termination criteria.
12. LHDs are widely spread not because they are superior, but because they are easy to implement and the underlying design principles are comprehensible. Only seven words are necessary to explain the design principle: "Place eight nonattacking castles on a chessboard."

5.9 Further Reading

Chapter 12 in Law & Kelton (2000) provides an introduction to the use of classical DOE techniques for computer simulations. Box et al. (1978) is a classical text on experimental design. Santner et al. (2003) give a survey of designs for modern DACE methods. Giunta et al. (2003) and Simpson et al. (2004) discuss different design considerations.

6

Search Algorithms

The alchemists in their search for gold discovered many other things of greater value.
—Arthur Schopenhauer

This chapter describes search algorithms for unconstrained optimization. The focus lies on the determination of their exogenous strategy parameters (design variables) to define the associated algorithm design. A short description of these algorithms is given, too.

We distinguish deterministic from stochastic search algorithms. Methods that can be found in standard books on continuous optimization such as Nocedal & Wright (1999) are characterized here as deterministic optimization algorithms. Stochastic or random strategies can be defined as methods "in which the parameters are varied according to probabilistic instead of deterministic rule"(Schwefel 1995, p. 87). Hoos & Stützle (2005) use the term "stochastic local search" for algorithms that generate or select candidate solutions randomly. Stochastic local search algorithms are popular for combinatorial optimization problems. If the function is continuous in its first derivative, gradient methods are usually more efficient than direct methods. *Direct methods* use only function evaluations. There are deterministic algorithms, for example, the simplex search of Nelder and Mead, and stochastic direct search algorithms, for example, evolution strategies.

6.1 Deterministic Optimization Algorithms

6.1.1 Nelder and Mead

The *Nelder–Mead simplex* (NMS) algorithm was motivated by the observation that $(d + 1)$ points are adequate to identify a downhill direction in a d-dimensional landscape (Nelder & Mead 1965). However, $(d+1)$ points define also a nondegenerated simplex in $\mathbb{R}^d$. Thus, it seemed a good idea to exploit a simplex for probing the search space, using only function values (Lewis et al. 2000).

Nelder and Mead incorporated a set of moves that enhance the algorithm's performance, namely *reflection*, *expansion*, *contraction*, and *shrinkage*. A new point is generated at each iteration. Its function value is compared to the function values at the vertices of the simplex. One of the vertices is replaced by the new point. Reflection reflects a vertex of the simplex through the centroid of the opposite face. Expansion allows the algorithm to take a longer step from the reflection point (centroid) toward the reflected vertex, while contraction halves the length of the step, thereby resulting in a more conservative search. Finally, shrinkage reduces the length of all edges that are adjacent to the best vertex, i.e., the vertex with the smallest function value. Thus, there are four design variables to be specified, namely the coefficients of reflection ρ, expansion χ, contraction γ, and shrinkage σ. Default settings of these parameters are reported in Table 6.1. NMS is considered to be quite a robust but relatively slow algorithm that works reasonably well even for nondifferentiable functions (Lagarias et al. 1998).

Table 6.1. Default settings (algorithm design) of the exogenous parameters (design variables) of the NMS algorithm. This design is used in the MATLAB optimization toolbox (Lagarias et al. 1998)

Symbol	Parameter	Range	Default
ρ	Reflection	$\rho > 0$	1.0
χ	Expansion	$\chi > \max\{1, \rho\}$	2.0
γ	Contraction	$0 < \gamma < 1$	0.5
σ	Shrinkage	$0 < \sigma < 1$	0.5

The MATLAB function `fminsearch` was used to perform the experiments. It uses the following values for the design variables: $\rho = 1$, $\chi = 2$, $\gamma = 0.5$, $\sigma = 0.5$.

6.1.2 Variable Metric

The variable metric method is a quasi-Newton method. Quasi-Newton methods build up curvature information. Let H denote the Hessian, c a constant vector, and b a constant, then a quadratic model problem formulation of the form

$$\min_x \frac{1}{2} x^T H x + c^T + b$$

is constructed. If the partial derivatives of x go to zero, that is,

$$\nabla f(x^*) = Hx^* + c = 0,$$

the optimal solution for the quadratic problem occurs. Hence

$$x^* = -H^{-1}c.$$

Quasi-Newton methods avoid the numerical computation of the inverse Hessian H^{-1} by using information from function values $f(x)$ and gradients $\nabla f(x)$

to build up a picture of the surface to be optimized. The MATLAB function `fminunc` uses the formula of Broyden (1970), Fletcher (1970), Goldfarb (1970), and Shanno (1970) to approximate H^{-1}. The gradient information is derived by partial derivatives using a numerical differentiation via finite differences. A line search is performed at each iteration in the direction

$$-H_k^{-1} \cdot \nabla f(x_k).$$

6.2 Stochastic Search Algorithms

6.2.1 The Two-Membered Evolution Strategy

The two-membered evolution strategy, or $(1+1)$-ES, is included in our analysis for three reasons: (i) It is easy to implement, (ii) it requires only a few exogenous parameters, and (iii) it defines a standard for comparisons. Many optimization practitioners apply the $(1+1)$-ES to their optimization problem. Schwefel (1995) describes this algorithm as "the minimal concept for an imitation of organic evolution." Let f denote an objective function $f : \mathbb{R}^d \to \mathbb{R}$ to be minimized. The rules of a $(1+1)$-ES can be described as shown in Figure 6.1. The standard deviation σ will be referred to as *step width* or *mutation strength*. The standard deviation σ is interpreted as the mean *step length*. The ratio of the number of the successes to the total number of mutations, the so-called *success rate* $1/s_r$, might be modified as well as the factor by which the variance is reduced or increased, the so-called *step-size adjustment factor* s_a. A related algorithm design is shown in Table 6.2.

Figure 6.2 shows the $1/5$ success rule derived by Rechenberg while analyzing the $(1+1)$-ES on two basically different objective functions for selecting appropriate step lengths (Rechenberg 1973).

A more precise formulation is required to implement the $1/5$ success rule. "From time to time during the optimization run" can be interpreted as "after every s_u mutations." The ratio of the number of the successes to the total number of mutations, the so-called *success rate* $1/s_r$, might be modified as well as the factor by which the variance is reduced or increased, the so-called *step-size adjustment* factor s_a. The number of iterations s_n to estimate the success rate has to be specified. Other schemes to modify the variance are possible, e.g., to additive or exponential variations. Furthermore, a *starting value for the step size* $\sigma^{(0)}$ has to be specified. The algorithm design for the $(1+1)$-ES is summarized in Table 6.2.

A coding of the two-membered ES and an in-depth discussion of evolution strategies and other direct search methods can be found in Schwefel's seminal book *Evolution and Optimum Seeking* (Schwefel 1995). This book is a slightly extended version of Schwefel's doctoral thesis from 1975 that was published under the title *Numerische Optimierung von Computer-Modellen mittels der Evolutionsstrategie* (Schwefel 1977), and translated into English four years later (Schwefel 1981).

Procedure: $(1+1)$-ES.

Initialization: Initialize the iteration counter: $t = 1$. Determine: (i) a point $X_1^{(t)}$ with associated position vector $x_1^{(t)} \in \mathbb{R}^d$ and (ii) a standard deviation $\sigma^{(t)}$. Determine the function value $y_1 = f(x_1^{(t)})$.

`while` some stopping criterion is not fulfilled `do`

Mutation: Generate a new point $X_2^{(t)}$ with associated position vector $x_2^{(t)}$ as follows:

$$x_2^{(t)} = x_1^{(t)} + z, \tag{6.1}$$

where z is a d-dimensional vector. Each component of z is the realization of a normal random variable Z with mean zero and standard deviation $\sigma^{(t)}$.

Evaluation: Determine the function value $y_2 = f(x_2^{(t)})$.

Selection: Accept $X_2^{(t)}$ as $X_1^{(t+1)}$ if

$$y_2 < y_1, \tag{6.2}$$

otherwise retain $X_1^{(t)}$ as $X_1^{(t+1)}$. Increment t.

Adaptation:

$$\text{Update} \quad \sigma^{(t)}. \tag{6.3}$$

`done.`

Fig. 6.1. The two-membered evolution strategy or $(1+1)$-ES for real-valued search spaces. The symbol f denotes an objective function $f : \mathbb{R}^d \to \mathbb{R}$ to be minimized

6.2.2 Multimembered Evolution Strategies

An ES-algorithm run can be described briefly as follows: The parental population is *initialized* at time (generation) $g = 0$. Then λ offspring individuals are generated in the following manner: A parent family of size ρ is selected randomly from the parent population. *Recombination* is applied to the object variables and the strategy parameters. The mutation operator is applied to the resulting offspring vector. After evaluation, a *selection* procedure is performed to determine the next parent population. The populations created in

1/5 Success Rule: From time to time during the optimization obtain the frequency of successes, i.e., the ratio of the number of the successes to the total number of trials (mutations). If the ratio is greater than 1/5, increase the variance; if it is less than 1/5, decrease the variance.

Fig. 6.2. Heuristic rule: 1/5 Success Rule

Table 6.2. Factors of the two-membered evolution strategy. Based on the default values, the step size σ is multiplied by 0.85 if the success rate is larger than $1/s_r = 1/5$, or equivalently, if more than 20 out of 100 mutations have been successful. The symbol d denotes the problem dimension

Symbol	Factor	Range	Default
s_n	Adaptation interval	$\mathbb{N}$	$10d$
s_u	Update interval	$\mathbb{N}$	d
s_r	1/success rate	$\mathbb{R}_+$	5
s_a	Step-size adjustment factor	$\mathbb{R}_+$	0.85
$\sigma^{(0)}$	Starting value of the step size σ	$\mathbb{R}_+$	1

the iterations of the algorithm are called *generations* or *reproduction cycles*. A termination criterion is tested. If this criterion is not fulfilled, the generation counter (g) is incremented and the process continues with the generation of the next offspring.

We consider the parameters or control variables from Table 6.3. This table shows typical parameter settings. Bäck (1996) presents a kind of default hierarchy that includes four parameterizations for simple and complex algorithms and suggests to perform experiments. Hence, our approach can be seen as an extension of Bäck's methods.

The reader is referred to Bartz-Beielstein (2003) for a detailed description of these parameters. Schwefel et al. (1995) and Beyer & Schwefel (2002) provide a comprehensive introduction to this special class of EA.

Table 6.3. Default settings of exogenous parameters of a "standard" evolution strategy (Bäck 1996). Bäck does not recommend using "standard" without reflection. Problems may occur when these "standards" are blindly adopted and not adjusted to the specific optimization problem. The offspring-parent ratio is defined as $\nu = \lambda/\mu$, d is the problem dimension, and r_d and r_i denote discrete and intermediary recombination, respectively. It is a common practice to choose 1 or d different standard deviations and a value of 2 or μ for the mixing number. Therefore n_σ, ρ, r_x, r_σ, and κ are treated as qualitative factors

Symbol	Parameter	Range	Default
μ	Number of parent individuals	$\mathbb{N}$	15
ν	Offspring–parent ratio	$\mathbb{R}_+$	7
$\sigma_i^{(0)}$	Initial standard deviations	$\mathbb{R}_+$	3
n_σ	Number of standard deviations	$\{1, 2, \ldots, d\}$	1
c_τ	Multiplier for the learning rate	$\mathbb{R}_+$	1
ρ	Mixing number	$\{1, 2, \ldots, \mu\}$	2
r_x	Recombination operator for object variables	$\{r_i, r_d\}$	r_d
r_σ	Recombination operator for strategy variables	$\{r_i, r_d\}$	r_i
κ	Maximum age	$\mathbb{N}$	1

6.2.3 Particle Swarm Optimization

The flocking behavior of swarms and fish shoals was the main inspiration which led to the development of particle swarm optimization algorithms (Kennedy & Eberhart 1995). PSO belongs to the class of stochastic, population-based optimization algorithms. It exploits a population of individuals to probe the search space. In this context, the population is called a *swarm* and the individuals are called *particles*. Each particle moves with an adaptable velocity within the search space, and it retains in a memory the best position it has ever visited. PSO has been applied to numerous simulation and optimization problems in science and engineering (Kennedy & Eberhart 2001; Parsopoulos & Vrahatis 2002, 2004). PSO's convergence is controlled by a set of design variables that are usually either determined empirically or set equal to widely used default values.

There are two main variants of PSO with respect to the information exchange scheme among the particles. In the *global* variant, the best position ever attained by all individuals of the swarm is communicated to all the particles at each iteration. In the *local* variant, each particle is assigned to a neighborhood consisting of prespecified particles. In this case, the best position ever attained by the particles that comprise a neighborhood is communicated among them. Neighboring particles are determined based on their indices rather than on their actual distance in the search space. Clearly, the global variant can be considered as a generalization of the local variant, where the whole swarm is considered as the neighborhood for each particle. In the current work we look at the global variant only.

Assume a d-dimensional search space, $S \subseteq \mathbb{R}^d$, and a swarm consisting of s particles. The ith particle is a d-dimensional vector,

$$x_i = (x_{i1}, x_{i2}, \dots, x_{id})^T \in S.$$

The velocity of this particle is also a d-dimensional vector,

$$v_i = (v_{i1}, v_{i2}, \dots, v_{id})^T.$$

The best previous position encountered by the ith particle (i.e., its memory) in S is denoted by

$$p_i^* = (p_{i1}^*, p_{i2}^*, \dots, p_{id}^*)^T \in S.$$

Assume b to be the index of the particle that attained the best previous position among all the particles in the swarm, and t to be the iteration counter.

Particle Swarm Optimization with Inertia Weights

Then, the resulting equations for the manipulation of the swarm are (Eberhart & Shi 1998),

$$v_i(t+1) = wv_i(t) + c_1 r_1 (p_i^*(t) - x_i(t)) + c_2 r_2 (p_b^*(t) - x_i(t)), \quad (6.4)$$
$$x_i(t+1) = x_i(t) + v_i(t+1), \quad (6.5)$$

where $i = 1, 2, \ldots, s$; w is a parameter called the *inertia weight*; c_1 and c_2 are positive constants, called the *cognitive* and *social* parameter, respectively; and r_1, r_2 are vectors with components uniformly distributed in $[0, 1]$. All vector operations are performed componentwise.

Usually, the components of x_i and v_i are bounded as follows:

$$x_{\min} \leqslant x_{ij} \leqslant x_{\max}, \quad -v_{\max} \leqslant v_{ij} \leqslant v_{\max}, \quad j = 1, \ldots, n, \quad (6.6)$$

where $x_{\min}$ and $x_{\max}$ define the bounds of the search space, and $v_{\max}$ is a parameter that was introduced in early PSO versions to avoid swarm explosion, which was caused by the lack of a mechanism for controlling the velocity's magnitude. Although the inertia weight is such a mechanism, empirical results have shown that using $v_{\max}$ can further enhance the algorithm's performance. Table 6.4 summarizes the design variables of particle swarm optimization algorithms.

Experimental results indicate that it is preferable to initialize the inertia weight with a large value, in order to promote global exploration of the search space, and gradually decrease it to get more refined solutions. Thus, an initial value around 1 and a gradual decline toward 0 is considered a proper choice for w. This scaling procedure requires the specification of the maximum number of iterations $t_{\max}$. Bartz-Beielstein et al. (2004a) illustrate a typical implementation of this scaling procedure.

Proper fine-tuning of the parameters may result in faster convergence and alleviation of local minima (Bartz-Beielstein et al. 2004a; Eberhart & Shi 1998; Beielstein et al. 2002b; Bartz-Beielstein et al. 2004b). Different PSO versions, such as PSO with constriction factor, have been proposed (Clerc & Kennedy 2002).

Table 6.4. Default algorithm design $x_{\mathrm{PSO}}^{(0)}$ of the PSO algorithm. Similar designs were used in Shi & Eberhart (1999) to optimize well-known benchmark functions

Symbol	Parameter	Range	Default	Constriction
s	Swarm size	$\mathbb{N}$	40	40
c_1	Cognitive parameter	$\mathbb{R}_+$	2	1.494
c_2	Social parameter	$\mathbb{R}_+$	2	1.494
$w_{\max}$	Starting value of the inertia weight w	$\mathbb{R}_+$	0.9	0.729
w_{scale}	Final value of w in percentage of $w_{\max}$	$\mathbb{R}_+$	0.4	1.0
$w_{\mathrm{iterScale}}$	Percentage of iterations, for which $w_{\max}$ is reduced	$\mathbb{R}_+$	1.0	0.0
$v_{\max}$	Maximum value of the step size (velocity)	$\mathbb{R}_+$	100	100

Particle Swarm Optimization with Constriction Coefficient

In the *constriction factor variant,* Eq. (6.4) reads,

$$v_i(t+1) = \chi\left[v_i(t) + c_1 r_1\left(p_i^*(t) - x_i(t)\right) + c_2 r_2\left(p_b^*(t) - x_i(t)\right)\right], \quad (6.7)$$

where χ is the *constriction factor* (Kennedy 2003). Equations (6.4) and (6.7) are related.

In our experiments, the so-called *canonical PSO* variant proposed in Kennedy (2003), which is the constriction variant defined by Eq. (6.7) with $c_1 = c_2$, has been used. The corresponding parameter setting for the constriction factor variant of PSO is reported in the last column (denoted as "Constriction") of Table 6.4, where χ is reported in terms of its equivalent inertia weight notation, for uniformity reason. Table 6.5 sumarizes the design variables of PSO with constriction factor. Shi (2004) gives an overview of current PSO variants.

Table 6.5. Default settings of the exogenous parameters of PSO with constriction factor. Recommendations from Clerc & Kennedy (2002)

Symbol	Parameter	Range	Default
s	Swarm size	$\mathbb{N}$	40
χ	Constriction coefficient	$\mathbb{R}_+$	0.729
φ	Multiplier for random numbers	$\mathbb{R}_+$	4.1
$v_{\max}$	Maximum value of the step size (velocity)	$\mathbb{R}_+$	100

6.3 Summary

The ideas presented in this chapter can be summarized as follows:

1. An algorithm design consists of one or more parameterizations of an algorithm. It describes exogenous strategy parameters that have to be determined before the algorithm is executed.
2. The MATLAB function `fminunc`, which implements a quasi-Newton method, has been presented as an algorithm that can be run without specifying exogenous strategy parameters.
3. Exogenous strategy parameters have been introduced for the following stochastic and deterministic optimization algorithms:
   (a) Nelder–Mead simplex algorithm
   (b) Evolution strategies (two-membered and multimembered versions)
   (c) Particle swarm optimization (inertia weight and constriction factor versions)

6.4 Further Reading

Nocedal & Wright (1999) give a good introduction to numerical optimization. Lagarias et al. (1998) discuss convergence properties of the Nelder–Mead simplex method. Press et al. (1992) can be used as a cookbook, because it is easy to read and presents many examples. Schwefel (1995) is the reference for ES, which covers classical algorithms as well. Rudolph (1997a) provides a deep analysis of convergence properties of evolutionary algorithms. Kennedy & Eberhart (2001) is a standard textbook on particle swarm optimization. Hoos & Stützle (2005) present a comprehensive overview of stochastic local search algorithms for combinatorial optimization problems.

Part II

Results and Perspectives

7

Comparison

> What is man in nature? A nothing in comparison with the infinite, an all in comparison with the nothing—a mean between nothing and everything.
> —Blaise Pascal

In Sect. 5.1 a distinction was drawn between *endogenous* and *exogenous* algorithm parameters. Exogenous parameters must be specified before the algorithm is started, but endogenous parameters can evolve during the optimization process, e.g., in self-adaptive evolution strategies (Beyer & Schwefel 2002). Usually, the adaptation of endogenous parameters depends on exogenous parameters. By varying the values of the exogenous parameters the experimenter can get some insight into the behavior of an algorithm.

Exogenous parameters will be referred to as *design variables* in the context of statistical design and analysis of experiments. The parameter values chosen for the experiments constitute an algorithm design X_A as introduced in Sect. 5.1. A design point $x_a \in \mathcal{D}_A$ represents one specific parameter setting.

> Algorithm *tuning* can be understood as the *process of finding the optimal* design point $x_a^* \in \mathcal{D}_A$ for a given problem design X_P.

The tuning procedure leads to results that are tailored for one specific algorithm-optimization problem combination. We cannot discuss the behavior of an algorithm without taking the underlying problem into account. A problem being PSO easy may be ES hard, and vice versa. The interaction between parameterizations of algorithms and problem difficulties has been discussed by other authors, Naudts & Kallel (2000) mention "the nonsense of speaking of a problem complexity without considering the parameterization of the optimization algorithm."

Tuning enables a fair comparison of two or more algorithms that should be performed prior to their comparison. This should provide an equivalent budget—for example, a number of function evaluations or an overall run time—for each algorithm.

It is crucial to formulate the goal of the tuning experiments precisely. Tuning was introduced as an optimization process. However, in many real-world situations, it is not possible or not desired to find the optimum. Assumptions, or *boundary conditions*, that are necessary for optimization have been analyzed in operations research (OR). These assumptions comprise conditions such as (1) well-defined goals, (2) stable situations and decision maker's values, or (3) an exhaustive number of alternatives. The review of these conditions demonstrates that "outside the limited-context problems presented in laboratory studies" (Klein 2002, p.113) only very few decision problems permit optimization. The so-called *fiction of optimization* is discussed in Sect. 7.1. Progressive deepening, a strategy used by chess grandmasters that is described in de Groot (1978), can be used as a "vehicle for learning," whereas decision analysis is a vehicle for calculating. This classification resembles Mayo's differentiation between NPT and NPT*. The final decision on whether a solution is optimal is similar to the final decision in a statistical test. Conditions required to make an optimal choice are considered in Sect. 7.2. This discussion is also relevant for the specification of *performance measures* (PM) in evolutionary computation. There are many different measures for the goodness of an algorithm, i.e., the quality of the best solution, the percentage of runs terminated successfully, or the number of iterations required to obtain the results.

Section 7.3 demonstrates how the classical DOE approach can be used to tune algorithms. It consists of three steps: screening, modeling, and optimization.

The modern DACE approach is presented in Sect. 7.4. This approach assumes that the correlation between errors is related to the distance between the sampling points, whereas linear regression used in DOE assumes that the errors are independent (Jones et al. 1998). Both approaches require different designs and pose different questions. In classical DOE, it is assumed that the data come from sources that are disturbed by random error. Designs with two or three levels only for each factor are used to build the corresponding regression models. DACE methods employ designs that vary each factor over many levels.

7.1 The Fiction of Optimization

Algorithm tuning as introduced in this chapter is an optimization problem. Many researchers describe *optimization* as the attempt to select the option with the highest expected utility (maximization).

Optimization relies on a number of very restrictive assumptions. No serious researcher would claim that these assumptions will be met in any setting, "with the possible exception of the laboratory or casino"(Klein 2002). Uncertainty, limited time, and restricted financial resources are only some of the reasons that prevent the determination of an optimal solution. The construction of a model that considers all these uncertainties requires a huge complexity. The resulting model cannot be applied in practice. The reader may reconsider the discussion from Sect. 2.1: "As far as the laws of mathematics refer to reality, they are not certain; and as far as they are certain, they do not refer to reality" (Newman 1956).

But why, despite these obvious problems, does the mathematical formulation of optimization linger as a gold standard for many researchers? Klein notes that the agenda for researchers is dictated by the mathematical formulation of expected utility: "... to find ways to translate decisions into the appropriate formalism." Deviations from this concept of maximization are seen as defects that can be eliminated. "Because maximization is based on mathematical proofs, these theorems act as a bedrock."

Klein questions the value of expected utility for understanding decision making. Instead of presenting a definition of optimization, he mentions important objections against commonly used ideas related to optimizations that are also relevant for comparing algorithms.

1. The optimization process plays an important role, because optimization does not only refer to the outcome. It is important to "provide accurate and reliable inputs to the analysis."
2. It is not obvious whether the optimization refers to the absolute best, the best solution given the data provided, or the best solution given all data that can be provided.
3. Stopping rule: "If we try to consider every relevant factor, we may not finish the analysis in a finite amount of time."
4. Suboptimal strategies are sometimes preferred in the engineering community; they are more robust than the optimal solution.

To define a measure that judges the performance of an algorithm, certain assumptions (boundary conditions) have to be fulfilled. Following Klein (2002), we will discuss some boundary conditions that have been compiled by decision researchers.

Boundary Conditions

The first assumption requires the goals to be well defined and specified in quantitative terms. This assumption appears to be unproblematic, because a performance measure can easily be defined. It is not a problem to find some performance measure—but it is a problem to find an appropriate one. Many performance measures can be defined, for example, the average function value from n optimization runs, the minimum value from these runs, or the median. These measures will be discussed in Sect. 7.2.

Other criteria demand that the decision maker's values as well as the optimization situation must be stable. However, the decision maker might gain new insight into the problem during the optimization. The optimization goal might be redefined due to this enhanced knowledge.

Another criterion demands that the decision maker is restricted to selections between options. But a typical decision maker is not only passively executing the experiments. Learning, even by accident, may occur. New ideas for improved algorithms can come up.

One criterion requires that the number of alternatives generated must be exhaustive and that the options must be properly compared to each other. We cannot test every single algorithm configuration and every possible problem instance. Even worse, results from this overarching test would be worthless due to the NFL theorem. However, experimental design techniques such as DOE or DACE can be applied to setup experiments systematically and more efficiently than the commonly used one-factor-at-a-time designs.

Furthermore, it is important that the optimal choice can be selected without wasting disproportionate time and effort. This criterion is related to *Fredkin's paradox,* which will be discussed in Chap. 8. Yet, it is not obvious how many instances of problem P are necessary to demonstrate that algorithm A performs better than algorithm B. Is a test suite with 500 functions more convincing than one with 5 functions?

7.2 Performance Measures

As tuning and comparison of search algorithms can be conducted for many reasons, different performance measures are necessary. Often, the average response value from an algorithm run is optimized. But there are circumstances under which it is desirable to optimize the maximum value and not the average. For example, to guarantee good service for all waiting customers in an elevator system, the maximum waiting time has to be minimized. Otherwise, for some systems, it is more important to minimize the variance in the response than it is to minimize the average value. Therefore, it is a good idea to show a graph that plots the quality of the solution versus its variance. However, providing this additional information turns simple optimization problems into multicriteria optimization problems.

Example 7.1 (Mean, median, maximum, and minimum). Considering the mean function values in Fig. 7.1, one might conclude that threshold selection (TS) improves the performance of the (1+1)-ES, especially for high noise levels. Comparing the median values leads to a similar conclusion. And, the comparison of the maximum function values shows that threshold rejection might improve the performance, too. However, the situation changes completely if the minimum function values are compared. Surprisingly, no clear difference between the two algorithms can be detected. ■

Obviously, it is not trivial to find adequate performance measures. The performance measure under consideration should lead to a comparison that is well-defined, algorithmic, reproducible, and fair (Johnson 2002). Dolan & More (2002) discuss several shortcomings of commonly used approaches, i.e., the subjectivity related to the choice of a penalty value that is assigned to algorithms that failed to solve a problem. We will consider three optimization scenarios before we present measures that refer to efficiency and those that refer to effectivity.

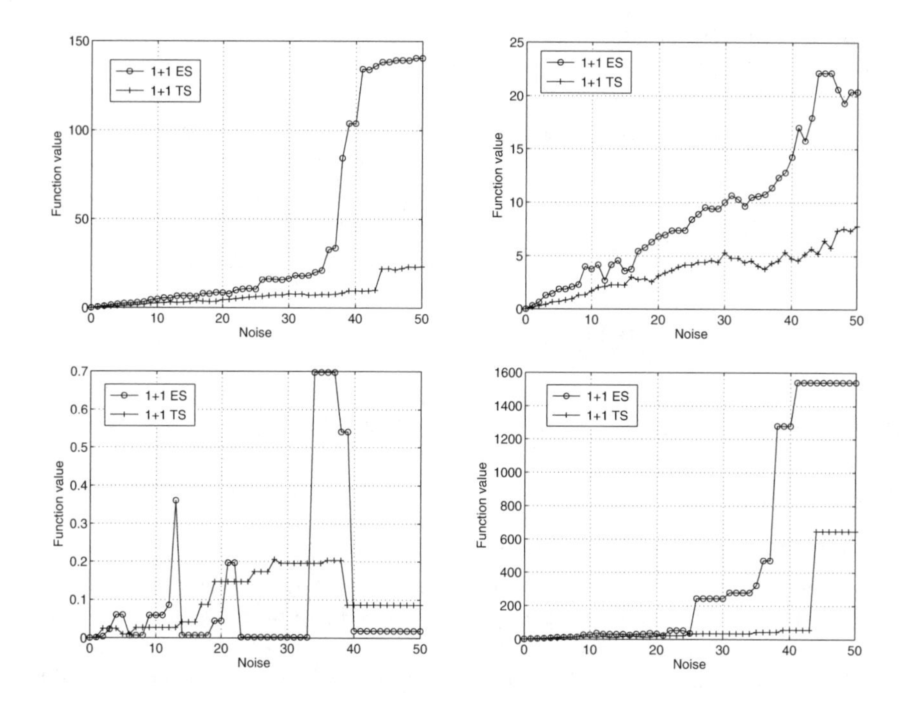

Fig. 7.1. Mean, median, maximum, and minimum function values. A $(1 + 1)$-ES compared to threshold selection (TS) for different noise strengths. Smaller values are better. From *left* to *right:* Mean and median (*first row*), and minimum and maximum function values (*second row*) from 500 runs for different noise levels

7.2.1 Scenarios

Several scenarios have been suggested in the literature: Rardin & Uzsoy (2001) distinguish research testing of new algorithms for existing problems from the development of the most efficient solution procedure for one problem instance.

Eiben & Smith (2003) differentiate between three types of optimization problems:

1. *Design problems* create one excellent solution at least once.
2. *Repetitive problems* find good solutions for different problem instances.
3. *Online control problems* are repetitive problems that have to be solved in real time.

Following Schwefel (1975, 1995), we will use a classification scheme that distinguishes between effectivity and efficiency. *Effectivity* is related to robustness and deals with the question whether the algorithm produces the desired effect. On the other hand, the measurement can be based on *efficiency*: Does the algorithm produce the desired effects without waste? The rate of convergence is one typical measure to judge the efficiency of evolutionary algorithms.

More et al. (1981) note that many tests do not place enough emphasis on testing the robustness of optimization programs. Most of the testing procedures are focused on their efficiency only. A notable exception in this context is Schwefel (1995), who performed three test series: the first to analyze numerically the rates of convergence for quadratic objective functions, the second to test the reliability of convergence for the general nonlinear case, and the third one to investigate the computational effort required for nonquadratic problems. Test functions for these scenarios were presented in Sect. 4.2.

7.2.2 Effectivity or Robustness

Robustness can be defined in many ways, i.e., as a good performance over a wide range of instances of one test problem or even over a wide range of different test problems. Criteria based on robustness mainly consider the best result. Robustness refers to the hardness or complexity of the problem. Therefore, the analysis can be based on low dimensional problems. Due to the lack of computing resources, Schwefel (1975) considered 50 problems with dimensions from 1 to 6 only. Machine precision demands the specification of a border f_{border} to distinguish solutions that have found a function value sufficiently close to the optimum value from solutions that failed to obtain this value. The *machine precision* ϵ is the largest positive number that $1 + \epsilon = 1$. For a computer that supports IEEE Standard 754, double-precision ϵ is $2^{-52} \approx 2.224 \times 10^{-16}$. A simplified variant of the border determination in Schwefel (1995, p. 206), reads as follows: Let $\epsilon \in \mathbb{R}_+$ be a real-valued positive constant, for example, the machine precision ϵ. Determine

$$f_{\text{border}} = \begin{cases} \max\{f(x^* + \epsilon x^*), f(x^* - \epsilon x^*)\}, & \text{if } x^* \neq 0, \\ \max\{f(\epsilon \mathbf{1}), f(-\epsilon \mathbf{1})\}, & \text{otherwise.} \end{cases}$$

We will list some commonly used performance measures to analyze the effectivity of an algorithm in the following:

(PM-7.1) If the optimal solution is known, the percentage of run configurations terminated successfully, the *success ratio* (SCR), can be used to measure the performance of an algorithm. The success ratio was already mentioned in Sect. 3.5 in the context of logistic regression models. Two variants of this measure can be defined: it can be based on the distance of the obtained best objective function value $\tilde{f}$ to the best known function value f^*, or on the distance of the position with the obtained best objective function value $\hat{x}$ to the position of best known function value x^*. Unless otherwise explicitly stated, we will use the variant that measures the distance between $\tilde{f}$ and f^*.

To measure the algorithm's progress toward a solution, one can specify a budget, i.e., the number of function evaluations available to an algorithm. If the starting points were chosen randomly, or if stochastic search algorithms were analyzed, several solutions are obtained for one algorithm configuration. We list only three possible ways to evaluate the results. Other measures, i.e., based on the median, are possible.

(PM-7.2) Schwefel (1995, p. 211) selects out of n tests the one with the best end result. Bartz-Beielstein (2005a) introduces a measure based on bootstrap, which reflects the goals of optimization practitioners to select the best results from several runs and to skip the others. The related algorithm is presented in Fig. 7.2.

(PM-7.3) The *mean best function value* can be defined as the average value of the best function values found at termination for one specific run configuration. This performance measure will be referred to as MBST.

(PM-7.4) The best function value found by an algorithm is recorded. By starting the algorithm from a number of randomly generated initial points, a sample is obtained. Trosset & Padula (2000) state that the construction of a nonparametric estimate of the probability density function from which the sample was drawn has an "enormous diagnostic value" to study the convergence of iterative algorithms to local solutions.

7.2.3 Efficiency

The methods presented in Sect. 7.2.2 specify the available resources in advance and ask how close to the optimum an algorithm could come. Diametrically opposed to these methods are those that measure the required resources. They measure the efficiency of an algorithm, for example, the number of function evaluations or the timing of the algorithm. This difference is depicted in Fig. 7.3.

(PM-7.5) Considering the quality of the best solution, it is a common practice to show a graph of the solution quality versus time. Due to the randomness of the results, it is useful to plot results from several runs. It is a good practice to add *error bars* to illustrate the confidence level of data or the deviation along a curve (Fig. 7.4) and not to show the mean values

Algorithm: Best out of n

1. Generate n results.
2. `repeat` k times:
   (a) Select (with replacement) a set M_i of $m < n$ values.
   (b) Determine $m_i := \min M_i$.
   `end.`
3. Calculate $\sum_i^k m_i/k$. The resulting value will be referred to as $\min_{\text{boot}}$.

Fig. 7.2. Algorithm to estimate the average of the best results from several runs

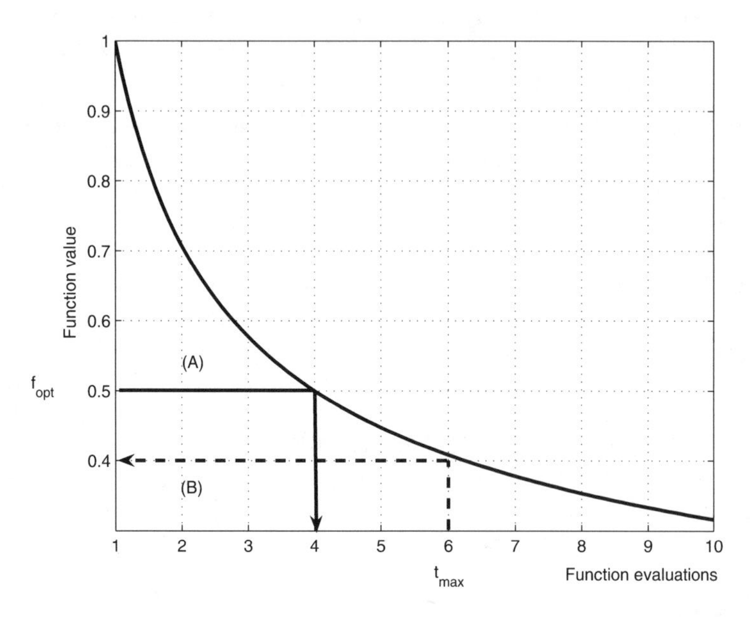

Fig. 7.3. Two diametrically opposed methods to determine the performance of an algorithm. *A* measures the required resources to reach a given goal, *B* measures the obtained function value that can be reached with a prespecified budget

only. Other techniques are possible, e.g., to add plots of the minimum and maximum values.

(PM-7.6) To measure the algorithm speed, the average number of evaluations to a solution can be used. The maximum number of evaluations can be used for runs finding no solutions.

(PM-7.7) The *run length distribution* (RLD) as introduced in Hoos (1998) provides suitable means to measure performance and to describe the qualitative behavior of optimization algorithms. RLDs are based on methods proposed in Parkes & Walser (1996).

A typical run length distribution is shown in Fig. 7.5. The algorithm to be analyzed is run n times with different seeds on a given problem instance. The maximum number of function evaluations $t_{\max}$ is set to a relatively high value. For each successful run the number of required function evaluations, t_{run}, is recorded. If the run fails, t_{run} is set to infinity. The empirical cumulative distribution represents these results. Let $t_{\text{run}}(j)$ be the run length for the jth successful run. Then, the empirical cumulative distribution is defined as

$$\Pr\left(t_{\text{run}}(j) \leqslant t\right) = \frac{\{\#j \,|\, t_{\text{run}}(j) \leqslant t\}}{n}, \tag{7.1}$$

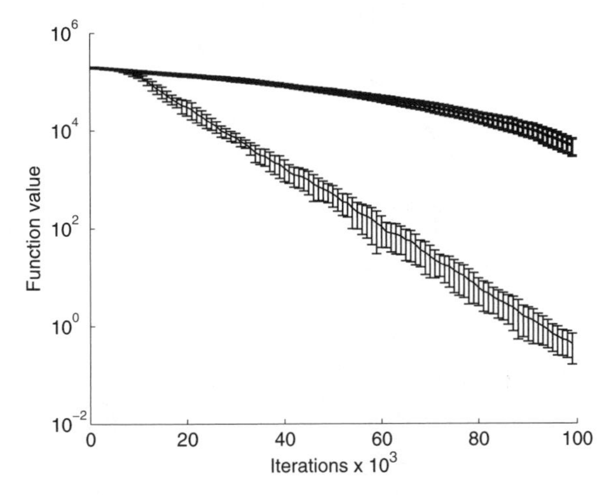

Fig. 7.4. Graph of the solution quality vs. time. Error bars added. The error bar is a distance of one standard deviation unit above and below the curve so that each bar is symmetric. Hence, error bars show the confidence level of data or the deviation along a curve

where $\{\#j \,|\, t_{\text{run}}(j) \leqslant t\}$ denotes the number of indices j, such that $t_{\text{run}}(j) \leqslant t$.

Exponential RLD can be used to determine whether a restart is advantageous. The exponential RLD is memoryless, because the probability of finding a solution within an interval $[t, t+k]$ does not depend on the actual iteration i. If the RLD is exponential, the number of random restarts does not affect the probability of finding a solution with a given interval. Otherwise, if the RLD is not exponential, there may exist some iteration for which a restart is beneficial. The reader is referred to the discussion in Chiarandini & Stützle (2002).

(PM-7.8) Efficiency rates measure progress from the starting point $x^{(0)}$ as opposed to convergence rates that use a point in the vicinity of the optimum x^*. Hillstrom (1977) defines the following efficiency measure:

$$\text{MTER} = \ln(|f^{(0)} - f^*|/|\hat{f} - f^*|)/T, \tag{7.2}$$

where T is an estimate of the elapsed time in centiseconds, and where $f^{(0)}$, f^*, and $\hat{f}$ are the initial, known, and final minimum values of the objective function. The difference $|\hat{f} - f^*|$ is bounded by the machine precision (Hillstrom 1977). Thus, Hillstrom's definition is not machine-independent.

(PM-7.9) A measure to compute the quality–effort relationship can be defined as the ratio $r_{0.05} = t_{0.05}/t_{\text{best}}$, where $t_{0.05}$ denotes the time to produce a solution within 5 % of the best function value found, and t_{best} is the time

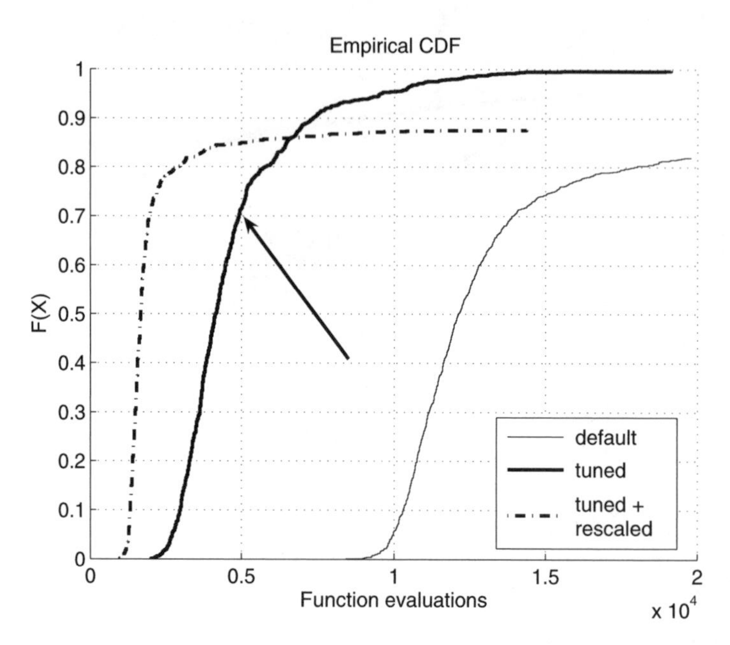

Fig. 7.5. Run length distributions of the default, the tuned, and the tuned and rescaled inertia variant of the particle swarm optimization. The *arrow* indicates that 70% of the runs of the tuned algorithm can detect the optimum with less than 5000 function evaluations. After 15,000 function evaluations nearly every run is able to detect the optimum. The $w_{\text{iterScale}}$ scheme is modified in the tuned and rescaled variant. It outperforms the other algorithms if the number of function evaluations is low. However, increasing the number of function evaluations from 5000 to 15,000 does not improve the success ratio of the tuned and rescaled PSO. The default configuration was able to complete only 80% of the runs after 20,000 function evaluations. RLDs provide good means to determine the maximum number of function evaluations for a comparison

to produce that best value (Barr et al. 1995). Run length distributions provide a graphical presentation of this relationship.

Example 7.2 (Quality–effort relationship). If an algorithm A requires $t^*(A) = 30,000$ function evaluations to produce the best solution $f^*(A) = 10$ and $t_{0.05}(A) = 10,000$ evaluations to produce a solution that is smaller than $f_{0.05}(A) = 10.5$, than its quality-effort relationship is 30,000/10,000 = 3. ■

Several measures have been defined for evolutionary algorithms. These measures have been introduced to measure the search progress, and not the convergence properties of an algorithm. For example, the corridor model with objective function

$$f(x) = -\sum_{i=1}^{d} x_i \tag{7.3}$$

and constraints as defined in Schwefel's Problem 3.8 (Schwefel 1995, p. 364) has its minimum at infinity. It was used to analyze "the cost, not of reaching a given approximation to an objective, but of covering a given distance along the corridor axis" (Schwefel 1995, p. 365). In particular, dynamic environments require algorithms that can follow the moving optimum, convergence is ipso facto impossible. Additionally, these measures consider that the best solution found during the complete run may not belong to the final population, for example, in comma-strategies:

(PM-7.10) Schwefel (1988) defines the *convergence velocity* c as a progress measure of a single run as the logarithmic square root of the ratio of the best function value in the beginning $f^{(0)}$ and after g generations $f^{(g)}$:

$$c(g) = \log(\sqrt{f^{(0)}/f^{(g)}}).$$

This measure is related to the efficiency measure MTER defined above in Eq. (7.2). Kursawe (1999) uses the normalized convergence velocity:

$$dc(g)/g,$$

where d denotes the problem dimension.

(PM-7.11) Rudolph (1997a) analyzes the first hitting time of an ϵ-environment of the global optimum of the objective function f^*.

(PM-7.12) Arnold & Beyer (2003) define the efficiency of a search algorithm as the ratio of the expected one-generation gain to the average number of objective function evaluations per generation.

Furthermore, we present three performance measures that are commonly used for evolutionary algorithms:

(PM-7.13) The quality gain $\overline{Q}$ measures the expected change of the function value from generation g to $g + 1$. It is defined as

$$\overline{Q} = E\left[1/s\sum_{i=1}^{s} f_i^{(g+1)} - 1/s\sum_{i=1}^{s} f_i^{(g)}\right], \tag{7.4}$$

where s denotes the population size.

(PM-7.14) The progress rate φ is a distance measure in the parameter space to measure the expected change in the distance of the parent population to the optimum x^* from generation g to $g+1$. This measure will be referred to as PRATE.

(PM-7.15) The success rate s_r is the ratio of the number of the successes to the total number of trials, i.e., mutations. It can be used as local performance measure from generation g to $g+1$. Note, the success rate was introduced in Sect. 6.2 to define the 1/5 success rule for evolution strategies.

The reader is referred to Beyer (2001) for a comprehensive discussion of these measures.

The performance measures considered so far are based on one problem instance only. The following measures compare i different problem instances on j different algorithm instances (Table 7.1).

(PM-7.16) For the ith problem instance and the jth algorithm, we can define

$$t_{i,j} = \text{time required to solve problem } i \text{ by algorithm } j. \tag{7.5}$$

The distribution function of a performance metric, the *performance profile*, shows important performance characteristics. It can be used to determine the computational effort (Dolan & More 2002; Bussieck et al. 2003). The *performance ratio* is defined as

$$r_{i,j} = \frac{t_{i,j}}{\min\{t_{i,j} : 1 \leq j \leq n_j\}}, \tag{7.6}$$

where n_j denotes the number of algorithms and $t_{i,j}$ is defined as in Eq. (7.5). The performance ratio $r_{i,j}$ compares the performance on problem i by algorithm j with the best performance by any algorithm on this problem. We can define

$$\rho_j(r) = \frac{1}{n_j} \#\{i : r_{i,j} \leq r\}, \tag{7.7}$$

the *cumulative distribution function (CDF) for the performance ratio.*

Example 7.3 (Performance ratio). Algorithm A_1 has the performance ratio on problem instance 1 from Table 7.1:

$$r_{1,1} = \frac{10}{\min\{10, 12, 5\}} = 2.$$

■

Example 7.4 (CDF for the performance ratio). Values of the cumulative distribution function for the performance ratio $\rho_j(r)$ can be determined as follows:

Table 7.1. Test and algorithm instances. The entries in the ith row and jth column present $t_{i,j}$, the number of function evaluations in units of 10^3 required to solve problem instance i with algorithm j as defined in Eq. (7.5). Smaller values are better

	Algorithm A_1	Algorithm A_2	Algorithm A_3
Problem 1	10	12	5
Problem 2	10	13	30
Problem 3	20	40	100

$$\rho_1(1) = 2/3, \text{ because there are 2 problem instances with } r_{i,1} \leq 1,$$
$$\rho_1(2) = 1, \text{ because there are 3 problem instances with } r_{i,1} \leq 2, and$$
$$\rho_1(3) = 1, \text{ because there are 3 problem instances with } r_{i,1} \leq 3.$$

The situation from Table 7.2 is depicted in Fig. 7.6. ■

Table 7.2. Performance ratios r_{ij} for the values from Table 7.1. Smaller values are better

	Algorithm A_1	Algorithm A_2	Algorithm A_3
Problem 1	2	2.4	1
Problem 2	1	1.3	3
Problem 3	1	2	5

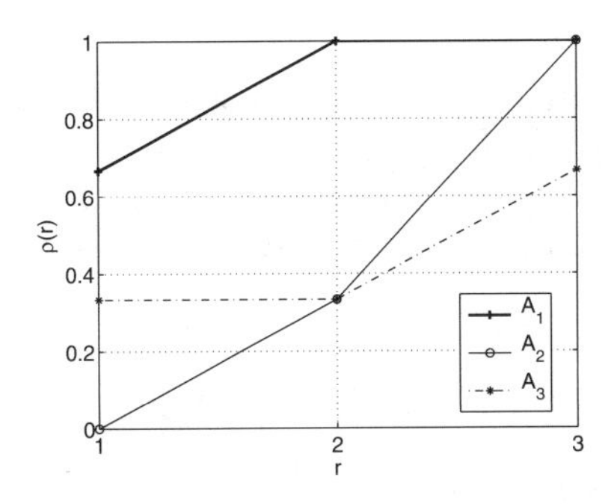

Fig. 7.6. Cumulative distribution function for the performance ratio $\rho_j(r)$. Values taken from Table 7.2. Algorithm A_1 performs best, whereas the performances of A_2 and A_3 cannot be distinguished. Note, larger values are better in this graph (in contrast to the values in Tables 7.1 and 7.2)

(PM-7.17) A common practice to compare global optimization algorithms (solvers) is to sort the problem instances by the time taken by a reference solver j_0: $t_i := t_{i,(j_0)}$. Then for every solver i the time taken $t_{i,j}$ is plotted against this ordered sequence of problem instances. Instances for which the optimum was not found by the solver within the allotted time $t_{\max}$ get a dummy time above the timeout value. The successful completion of the optimization task can be assessed directly. How the performance of a solver scales with the problem dimension can only be seen indirectly, since the values refer to the ordering of the problem instances induced by the reference solver (Neumaier et al. 2005).

(PM-7.18) Schwefel (1995, p. 180) suggests the following procedure to test the theoretical predictions of convergence rates: Test after each generation g, if the interval of uncertainty of the variables has been reduced by at least 90%:

$$|x_i^{(g)} - x_i^*| \leq \frac{1}{10}|x_i^{(0)} - x_i^*|, \quad \text{for } i = 1, \ldots, d, \tag{7.8}$$

where x^* denotes the optimum and the values x_i were initialized as in Eq. (5.6) in Example 5.5. The number of function calls can be displayed as a function of the numbers of parameters (dimension) on a log–log scale as shown in Fig. 7.7.

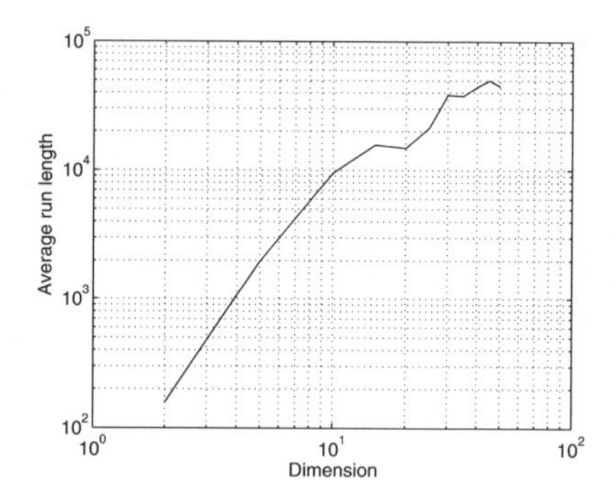

Fig. 7.7. Average run length vs. problem dimension. The number of function calls is displayed as a function of the numbers of parameters (dimension) on a log–log scale. A similar presentation was chosen in Schwefel (1995) to test convergence rates for a quadratic test function. The figure depicts data from an analysis of a particle swarm optimization on the sphere function

Since run times depend on the computer system, measures for computational effort might be advantageous: Counting operations, especially for major subtasks such as function calls, can explicitly be mentioned in this context. Finally, we note that Barr & Hickman (1993) discuss performance measures for parallel algorithms.

7.2.4 How to Determine the Maximum Number of Iterations

Measures based on effectivity often require the specification of $t_{\max}$, which is the maximum number of iterations before the run is started. A wrong specification of the $t_{\max}$ value may lead to a bad experimental design. Test problems that are too easy may cause ceiling effects. If algorithms A and B achieve the maximum level of performance (or close to it), the hypothesis "performance(A) $\geq$ performance(B)" should not be confirmed (Cohen 1995). Floor effects describe the same phenomenon on the opposite side of the performance scale: The test problem is too hard, so nearly no algorithm can solve it correctly.

Example 7.5 (Floor effects). If the number of function evaluations is chosen too small, floor effects can occur. Consider Rosenbrock's function, the starting point $x_i^{(0)} = (10^6, 10^6)^T$, and a budget of 10 function evaluations only. ■

Example 7.6 (Ceiling effects). The randomly generated instances of the TSP as described in Problem 4.1 can produce ceiling effects, because any random guess produces a good solution. ■

Run length distributions as presented in Sect. 7.2.3 can be used to determine an appropriate value for t_{max}.

Schaffer et al. (1989) propose a technique to determine the total number of iterations t_{max} and to prevent ceiling effects: The number k belongs to the set $\mathcal{N}$ if at least 10% of the algorithm design configurations $x_a \in \mathcal{D}_A$ located the known best objective function value f^* at least on average every second time after k iterations. The number of function evaluations at which to compare different algorithm designs is chosen as $N_{\text{tot}} = \min\{\mathcal{N}\}$.

Example 7.7 (Schaffer's technique to avoid ceiling effects). Based on the data from Table 7.1, we can see that algorithm 1 was able to locate the optimum at least every second time after 10,000 function evaluations, algorithms 2 and 3 required 13,000 and 30,000 evaluations, respectively, and $\mathcal{N} = \{10,000; 13,000; 30,000\}$. A choice of $k = 10,000$ as the minimum number of function evaluations guarantees that at least 10% of the algorithms locate the optimum after k iterations. At least 10% is in our case one algorithm only, because three algorithms are considered. ■

7.3 The Classical DOE Approach

Rather than detail the classical DOE procedure here, since it is fully outlined in Bartz-Beielstein (2003), we give an overview and make some comments that reflect further experiences that we have observed in the meantime.

7.3.1 A Three-Stage Approach

In classical DOE the three-stage approach of screening, modeling, and optimization is proposed:

(DOE-1) *Screening.* Consider an algorithm with k exogenous strategy parameters, for example, an evolution strategy with $k = 9$ parameters. Screening analyzes the main effects only. Possible interactions will be investigated later. Therefore, we recommend to use fractional-factorial 2^{k-p} designs with $1 \leq p < k$. These are orthogonal designs that require a moderate number of experiments. Due to the orthogonality of these designs, the regression coefficients can be determined independently. If we cannot differentiate between two effects, these effects are called *confounded*. A 2^{k-p} design is of resolution R if no q-factor effect is confounded with another effect that has less than $R - q$ factors (Box et al. 1978). Roman numerals denote the corresponding design resolution. Our first experiments are based on resolution III designs. These designs ensure that no main effect is confounded with any other main effect, but main effects can be confounded with two-factor interactions. Fractional-factorial 2^{k-p} designs provide unbiased estimators of the regression coefficients of a first-order model and can easily be augmented to designs that enable the estimation of a second-order regression model that will be used during optimization, the third stage of the classical DOE approach.

(DOE-2) *Modeling.* First- or second-order interactions can be taken into account, because only the important factors that have been detected during the screening phase will be analyzed further. At this stage, resolution IV or resolution V designs are recommended. Half-normal plots can be used to display the main effects and interactions. A linear approximation may be valid in a subdomain of the full experimental area. *Response surface methodology* (RSM) is a collection of mathematical and statistical tools to model, analyze, and optimize problems where the response of interest is influenced by several variables (Montgomery 2001). In RSM, we determine the direction of improvement using the "path of the steepest descent" (minimization problem) based on the estimated first-order model (Kleijnen & Van Groenendaal 1992). If no further improvement along the path of the steepest descent is possible, we can explore the area by fitting a local first-order model and obtain a new direction for the steepest descent. We can repeat this step until the expected optimum area is found (if the response surface is unimodal). There the linear model is inadequate and shows significant lack-of-fit and we cannot determine a direction of improved response in this case.

(DOE-3) *Optimization.* Central composite designs that can be complemented with additional axial runs are often used at this experimental stage. They can be combined with response surface methods and require a relatively high number of runs. We apply the standard techniques from regression analysis for metamodel validation (Draper & Smith 1998). A second-order model can be fitted in the expected optimum area. The optimal values are estimated by taking the derivatives of the second-order regression model. We combine in our approach DOE and RSM techniques that are adapted to the special needs and restrictions of the optimization task.

Example 7.8 (2^{3-1} designs). Consider a design for three factors A, B, and C with two levels each. One-half fraction of the full factorial 2^3 design is called a 2^{3-1} fractional-factorial design. Plus and minus signs can be used to denote high and low factor levels, respectively. If we define the multiplication of two factors by their associated levels as $++ = -- =: +$ and $+- = -+ =: -$, then one-half fraction of the design is given by selecting only those combinations with $ABC = +$, for example: $A = -$, $B = -$, and $C = +$. This design is a resolution III design. ■

7.3.2 Tuning an Evolution Strategy

Bartz-Beielstein (2003) describes a situation in which only a few preliminary experiments can be performed to find a suitable ES parameter setting. To start, an experimental region (design space) has to be determined. The design space is defined as

$$I := [a_1, b_1] \times \ldots \times [a_d, b_d] \subseteq \mathbb{R}^d, \tag{7.9}$$

with the center point $z_i = (a_i + b_i)/2, \quad i, \ldots, d$, as depicted in Fig. 5.1. The optimization response is approximated in the experimental region by the first-order regression model, cf. Eq. (3.12). The range of a coded or standardized variable x is bounded by $[-1, 1]$. The range $[a, b]$ of the corresponding original (natural) variable z can be mapped by a linear transformation to $[-1, 1]$.

An ES as presented in Beyer & Schwefel (2002) has at least nine different exogenous parameters. To model the ES performance, four quantitative variables $(\mu, \nu, \sigma^{(0)}, c_\tau)$ and five qualitative variables $(r_x, r_\sigma, \kappa, n_\sigma, \rho)$ have to be considered. The number of offspring λ can be determined from the size of the population μ and value of the selective pressure $\nu = \lambda/\mu$. The inputs μ and ν are treated as quantitative factors, their values are rounded to the nearest whole number to get a set of working parameters. The maximum lifespan κ is treated as a qualitative factor because only comma and plus selection schemes have been analyzed.

Instead of using a full factorial 2^k design that would require 512 optimization runs, a 2^{9-5}_{III} fractional-factorial design, which requires only 16 optimization runs, was chosen. Box et al. (1978) give rules for constructing fractional-factorial designs. First, the experimental region is chosen. The interval [−1Level, + 1Level] from Table 7.3 contains values that were proposed in Bäck (1996), cf. Table 6.3.

Example 7.9 (Experimental region). We have chosen the experimental region $I_1 = [10, 20]$ for μ, because it includes the recommended value $\mu = 15$. ■

Table 7.4 shows the 16 run configurations.

Table 7.3. Evolution strategy: symbols and levels. Values chosen with respect to the default settings from Table 6.3

Symbol	Parameter	Variable	Type	−1	+1
μ	Number of parent individuals	x_1	Quan	10	20
ν	Offspring–parent ratio	x_2	Quan	5	10
$\sigma^{(0)}$	Initial standard deviations	x_3	Quan	1	5
n_σ	Number of standard deviations	x_4	Qual.	1	12
c_τ	Multiplier for mutation parameters	x_5	Quan	1	2
ρ	Mixing number	x_6	Qual.	b	m
r_x	Recombination for object variables	x_7	Qual.	r_i	r_d
r_σ	Recombination for strategy variables	x_8	Qual.	r_i	r_d
κ	Maximum life span	x_9	Qual.	−1	1

Table 7.4. Fractional-factorial design for evolution strategies. Symbols were introduced in Table 6.3

	μ	ν	$\sigma^{(0)}$	n_σ	c_τ	ρ	r_x	r_σ	κ
1	10	5	1	1	1	2	i	i	1
2	20	5	1	1	2	2	d	d	-1
3	10	10	1	1	2	10	i	d	-1
4	20	10	1	1	1	20	d	i	1
5	10	5	5	1	2	10	d	i	-1
6	20	5	5	1	1	20	i	d	1
7	10	10	5	1	1	2	d	d	1
8	20	10	5	1	2	2	i	i	-1
9	10	5	1	12	1	10	d	d	-1
10	20	5	1	12	2	20	i	i	1
11	10	10	1	12	2	2	d	i	1
12	20	10	1	12	1	2	i	d	-1
13	10	5	5	12	2	2	d	d	1
14	20	5	5	12	1	2	i	i	-1
15	10	10	5	12	1	10	d	i	-1
16	20	10	5	12	2	20	i	d	1

A First Look at the Data

Box-plots, histograms, or scatterplots can be used to detect outliers easily. As randomness is replaced by pseudorandomness, we do not recommend simply excluding outliers from the analysis. Removing potential outliers may destroy valuable information. Instead, we recommend to look at the raw data that are tabulated and sorted. Specifying a better suited experimental region for factors that arouse suspicion might prevent outliers.

Example 7.10 (Outliers and experimental region). We can conclude from Table 7.5 that the choice of a value of 20 as the second level for factor A should be reconsidered. ■

Regression Analysis

Regression analysis and stepwise model selection by *Akaike's information criterion* have been performed for the coded variables x_i. To conduct the experiments, these values have be retransformed to the natural variables z_i. Before we start the search along the path of the steepest descent, the adequacy of the regression model is tested, and a check for interactions is performed. Regression analysis reveals that only three of the nine factors are important: (1) The initial sigma value $\sigma^{(0)}$, (2) the population size μ, (3) and the selective pressure ν. Plus and comma strategies are tested in parallel, because they perform similarly at this stage of experimentation.

Starting from the center point we perform a line search in the direction of the steepest descent that is given by $-(\hat{\beta}_1, \ldots, \hat{\beta}_k)$. To determine the step sizes Δx_i for the line search, we select the variable x_j that has the largest absolute regression coefficient: $j = \arg\max_i |\hat{\beta}_i|$. The increment in the other variables is

$$\Delta x_i = -\hat{\beta}_i / (|\hat{\beta}_j| / \Delta x_j), \qquad i = 1, 2, \ldots, k;\ i \neq j.$$

The corresponding numerical values are shown in Table 7.6. The qualitative factors have to be treated separately, because a line search cannot be performed for qualitative factors such as the recombination operator. For qualitative factors with significant effects the "better" levels were chosen. The values of qualitative factors with small effects on the response were chosen rather subjectively. Before the initial sigma value $\sigma^{(0)}$ reaches the boundaries of the feasible region, the search is stopped. This value may lie outside the experimental region, but has to be a feasible value for the algorithm. A sec-

Table 7.5. Tabulated raw data. The function value Y is shown in the first column. Factor A produces outliers, if its high level is chosen

Y	A	B	C	...
0.5	5	5	1	...
0.6	5	10	1	...
0.61	5	5	5	...
0.9	5	10	5	...
⋮	⋮	⋮	⋮	...
0.4	10	10	5	...
259.2	20	10	1	...
277.1	20	5	5	...
297.3	20	5	1	...
433.6	20	10	5	...

Table 7.6. Steepest descent; 12-dimensional sphere function. The line search is stopped after 7 steps to avoid negative $\sigma^{(0)}$ values (Bartz-Beielstein 2003)

Steps Δ	$\sigma^{(0)}$ Coded x_1	ν Coded x_2	μ Coded x_3	$\sigma^{(0)}$ Original z_1	ν Original z_2	μ Original z_3	Mean response $\log(y)$	Median response $\log(y)$
	0	0	0	3.0	8	15	−1.784	−1.856
Δ	−0.2	−0.15	−0.1	2.6	7	14	−3.169	−3.577
2Δ	−0.4	−0.3	−0.2	2.2	7	14	−3.184	−3.531
3Δ	−0.6	−0.45	−0.3	1.8	6	14	−4.231	−4.435
4Δ	−0.8	−0.6	−0.4	1.4	6	13	−5.018	−5.339
5Δ	−1.0	−0.75	−0.5	1.0	6	12	−6.445	−6.497
6Δ	−1.2	−0.9	−0.6	0.6	5	12	−7.451	−8.298
7Δ	−1.35	−1.05	−0.7	0.2	5	12	−8.359	−8.915

ond algorithm design based on the improved setting from the line search was created. The regression analysis shows that the plus selection scheme is advantageous in this case. Therefore, the experiments with the comma strategy, which have been run in parallel, are stopped. Effects caused by the parameters μ, ν, and $\sigma^{(0)}$ are statistically significant and have to be considered further. As only three of the nine parameters remain, a more complex algorithm design has been chosen.

Central Composite Designs

A central composite design combines a 2^k factorial design with additional runs, see Fig. 5.1. We can conclude from the regression analysis that was based on a 2^3 central composite design that a further decrease of the population size and of the selective pressure might be beneficial. The initial step size $\sigma^{(0)}$ has no significant effect any more. This result corresponds with the conclusions that might be drawn from the tree-based regression analysis (not shown here). Instead of performing a second line search, we use response surface methods to visualize the region of the local optimum. Data generated from a CCD with axial runs can be used to generate a surface plot. A numerical comparison of the function values obtained from the first and the improved algorithm design reveals a significant improvement (Table 7.7). We have found a better

Table 7.7. Evolution strategy. Comparison of the function values from the first and the improved algorithm design

Design	Min.	1st Qu.	Median	Mean	3rd Qu.	Max.
$x_{\text{ES}}^{(0)}$	0.01	0.04	0.05	0.07	0.09	0.20
x_{ES}^{*}	$5.32e-64$	$1.16e-59$	$1.25e-57$	$6.47e-51$	$7.54e-56$	$5.17e-49$

algorithm design, x_{ES}^{*}, that improves the performance of the "standard" ES presented in Bäck (1996) for this specific problem. This result is not surprising, since this standard was chosen as a "good" parameterization on average for many problems and not especially for the sphere model. The result found so far does not justify the conclusion that this design is optimal. Our intention is to give the optimization practitioner a framework on how to set up algorithms with working parameter configurations. Further optimization of this setting is possible, but this is beyond the intention of this study.

As noted in Sect. 5.6, the assumption of a linear model for the analysis of computer algorithms is highly speculative. The applicability of methods from computational statistics that are not restricted to this assumption is analyzed in the following section.

7.4 Design and Analysis of Computer Experiments

The term *computational statistics* subsumes computationally intensive methods (Gentle et al. 2004b). Statistical methods, such as experimental design techniques and regression analysis, can be used to analyze the experimental setting of algorithms on specific test problems. One important goal in the analysis of search algorithms is to find variables that have a significant influence on the algorithm's performance. Performance measures were discussed in Sect. 7.2, i.e., performance can be quantitatively defined as the average obtained function value in a number (e.g., 50) of independent experiments. This measure was also used in Shi & Eberhart (1999). Questions like "How does a variation of the swarm size influence the algorithm's performance?" or "Are there any interactions between swarm size and the value of the inertia weight?" are important research questions that provide an understanding of the fundamental principles of stochastic search algorithms such as PSO.

The approach presented in this section combines DOE, CART, and DACE techniques. Since DACE was introduced for deterministic computer experiments, repeated runs are necessary to apply this technique to stochastic search algorithms.

In the following, the specification of the DACE process model that will be used later to analyze our experiments is described. This specification is similar to the selection of a linear or quadratic regression model in classical regression. DACE provides methods to predict unknown values of a stochastic process, and it can be applied to interpolate observations from computationally expensive simulations. Furthermore, it enables the estimation of the prediction error of an untried point, or the mean squared error of the predictor.

Sequential Designs Based on DACE

Prior to the execution of experiments with an algorithm, the experimenter has to specify suitable parameter settings for the algorithm, i.e., a design point x_a from an algorithm design X_A.

Often, designs that use sequential sampling are more efficient than designs with fixed sample sizes. First, an initial design $X_A^{(0)}$ is specified. Information obtained in the first runs can be used for the determination of the second design $X_A^{(1)}$ in order to choose new design points more efficiently.

Sequential sampling approaches with adaptation have been proposed for DACE. For example, in Sacks et al. (1989) sequential sampling approaches with and without adaptation were classified to the existing metamodel. We will present a sequential approach that is based on the expected improvement. In Santner et al. (2003, p. 178) a heuristic algorithm for unconstrained global minimization problems is presented. Consider one problem design $x_p \in \mathcal{D}_P$. Let $y^k_{\min}$ denote the smallest known minimum value after k runs of the algorithm, $y(x)$ be the algorithm's response, i.e., the realization of $Y(x)$ in

Eq. (3.17), and let x_a represent a specific design point from the algorithm design X_A. Then the improvement is defined as

$$\text{improvement at } x_a = \begin{cases} y^n_{\text{min}} - y(x_a), & y^n_{\text{min}} - y(x_a) > 0, \\ 0, & \text{otherwise,} \end{cases}$$

for $x_a \in \mathcal{D}_A$, cf. Eq. (5.5). Combining this sequential approach with classical and further modern statistical methods (e.g., regression trees and bootstrapping) leads to the sequential parameter optimization that will be introduced in the following section.

7.5 Sequential Parameter Optimization

The *sequential parameter optimization* (SPO) method, which is developed in this section, describes an implementable but heuristic method.

During the *preexperimental planning phase* (S-1) the experimenter defines exactly what is to be studied and how the data are to be collected. The recognition and statement of the problem seems to be a rather obvious task. However, in practice, it is not simple to formulate a generally accepted goal. *Discovery*, *comparison*, *conjecture* and *robustness* as introduced in Sect. 2.1.2 are only four possible scientific goals of an experiment. Furthermore, the experimenter should take the boundary conditions discussed in Sect. 7.1 into account. Statistical methods like run length distributions provide suitable means to measure the performance and describe the qualitative behavior of optimization algorithms.

In step (S-2), the experimental goal should be formulated as a scientific claim, e.g., "Algorithm A, which uses a swarm size s, that is proportional to the problem dimension d outperforms algorithms that use a constant swarm size."

A statistical hypothesis, such as "There is no difference in means comparing the performance of the two competing algorithms," is formulated in the step (S-3) that follows.

Step (S-4) requires at least the specification of

(a) an optimization problem
(b) constraints (for example, the maximum number of function evaluations)
(c) an initialization method
(d) a termination method
(e) an algorithm, and its important factors
(f) an initial experimental design
(g) a measure to judge the performance

Regarding (c), several methods have been used for the initialization of the population in population-based algorithms, or the determination of an initial point, $x^{(0)}$, in algorithms that use a single search point. For example, an

asymmetric initialization scheme was used in Shi & Eberhart (1999), where the initial positions of the particles, $x_i^{(0)}$, $i = 1, \dots, s$, were chosen uniformly distributed in the range $[15, 30]^d$. Initialization method DETMOD, which uses deterministically modified starting values, was proposed in More et al. (1981).
An algorithm terminates if the problem was solved (XSOL), the algorithm has stalled (STAL), or the resources, e.g., the maximum number of function evaluations, $t_{\max}$, are exhausted (EXH). Note that initialization and termination methods were discussed in Sect. 5.5.

The corresponding problem design X_P that summarizes the information from (a) to (d) for our experiments with PSO is reported in Table 7.8, while the algorithm design X_A, which represents (e), is reported in Table 7.9 . The experimental goal of the sequential approach presented here can be characterized as the determination of an optimal (improved) algorithm design point, x^*_{PSO} for a given problem design point $x^{(0)}_{\text{PSO}}$.

At each stage, Latin hypercube designs are used. Aslett et al. (1998) report that experience with the stochastic process model had indicated that 10 times the expected number of algorithm design variables is often an adequate number of runs for the initial LHD.

Example 7.11 (LHD for the PSO constriction variant). The constriction factor variant of PSO requires the determination of four exogenous strategy parameters, namely the swarm size s, constriction factor χ, param-

Table 7.8. Problem design $x^{(1)}_{\text{rosen}}$ for the experiments performed in this chapter. The experiment's name, the number of runs n, the maximum number of function evaluations $t_{\max}$, the problem's dimension d, the initialization method, the termination criterion, the lower and upper bounds, x_l and x_u, respectively, for the initialization of the object variables $x_i^{(0)}$, as well as the optimization problem and the performance measure (PM) are reported

Design	Init.	Term.	PM	n	$t_{\max}$	d	x_l	x_u
$x^{(1)}_{\text{rosen}}$	NUNIRND	EXH	MBST	50	2500	10	15	30

Table 7.9. PSO: Algorithm designs for the inertia weight PSO variant. They correspond to the experiment $x^{(1)}_{\text{rosen}}$ of Table 7.8, which optimizes the 10-dimensional Rosenbrock function. $x^{(l)}_{\text{PSO}}$ and $x^{(u)}_{\text{PSO}}$ denote the lower and upper bounds to generate the LHD, respectively, and x^*_{PSO} denotes the parameter settings of the improved design that was found by the sequential approach

Design	s	c_1	c_2	$w_{\max}$	w_{scale}	$w_{\text{iterScale}}$	$v_{\max}$
$x^{(l)}_{\text{PSO}}$	5	1.0	1.0	0.7	0.2	0.5	10
$x^{(u)}_{\text{PSO}}$	100	2.5	2.5	0.99	0.5	1	750
x^*_{PSO}	21	2.25	1.75	0.79	0.28	0.94	11.05

eter $\varphi = c_1 + c_2$, and the maximum velocity $v_{\max}$. Thus, an LHD with at least $m = 15$ design points was chosen. This is the minimum number of design points to fit a DACE model that consists of a second-order polynomial regression model and a Gaussian correlation function. The former requires $1 + \sum_{i=1}^{4} i = 11$ design points, while the latter requires 4 design points. Note that for $m = 15$ there are no degrees of freedom left to estimate the mean squared error of the predictor (Santner et al. 2003). ■

After that, the experiment is run (S-5). Preliminary (pilot) runs can give a rough estimate of the experimental error, run times, and the consistency of the experimental design. Again, RLDs can be very useful. Since we consider probabilistic search algorithms in our investigation, design points must be evaluated several times.

The experimental results provide the base for modeling and prediction in step (S-6). The model is fitted and a predictor is obtained for each response.

The model is evaluated in step (S-7). Several visualization techniques can be applied. Simple graphical methods from exploratory data analysis are often helpful. Histograms and scatterplots can be used to detect outliers. If the initial ranges for the designs were chosen improperly (e.g., very wide initial ranges), visualization of the predictor can guide the choice of more suitable (narrower) ranges in the next stage. Several techniques to assess the validity of the model have been proposed.

Additional graphical methods can be used to visualize the effects of factors and their interactions on the predictors. The three-dimensional visualizations depicted in Fig. 7.9, produced with the DACE toolbox (Lophaven et al. 2002b), have proved to be very useful. The predicted values can be plotted to support the numerical analysis, and the MSE of prediction is used to asses its accuracy. We explicitly note here that statistical models can provide only guidelines for further experiments. They do not prove that a factor has a particular effect.

If the predicted values are not accurate, the experimental setup has to be reconsidered. This includes the scientific goal, the ranges of the design variables, and the statistical model (Eqs. (3.12) and (3.17)). New design points in promising subregions of the search space can be determined (S-8) if further experiments are necessary.

Thus, a termination criterion has to be tested (S-9). If it is not fulfilled, based on the expected improvement defined in Eq. (5.5) new candidate design points can be generated (S-10). A new design point is selected if there is a high probability that the predicted output is below the current observed minimum and/or there is a large uncertainty in the predicted output. Otherwise, if the termination criterion is true, and the obtained solution is good enough, the final statistical evaluation (S-11) that summarizes the results is performed. A comparison between the first and the improved configuration should be performed. Techniques from exploratory data analysis can complement the analysis at this stage. Besides an investigation of the numerical values, such as mean, median, minimum, maximum, $\min_{\text{boot}}$ and standard deviation, graphical presentations such as boxplots, histograms, and RLDs can be used to support the final statistical decision (e.g., see Fig. 7.11).

Finally, we have to decide whether the result is scientifically important (S-12), since the difference, although statistically significant, can be scientifically meaningless. As discussed in Sect. 2.5, an objective interpretation of rejecting or accepting the hypothesis from (S-2) should be presented here. Consequences that arise from this decision are discussed as well. The experimenter's skill plays an important role at this stage. The experimental setup should be reconsidered at this stage and questions like "Have suitable test functions or performance measures been chosen?" or "Did floor or ceiling effects occur?" must be answered. Test problems that are too easy may cause such ceiling effects, cf. the discussion in Sect. 7.2.4.

7.6 Experimental Results

Initially, we investigated Rosenbrock's function. This is a simple and well-known test function to gain an intuition regarding the functioning of the proposed technique (Rosenbrock 1960). In the next step of our analysis, the S-ring model was considered. We provide a demonstration of the sequential approach by conducting a brief investigation for the Rosenbrock function, using the two variants of PSO as well as the Nelder–Mead simplex algorithm.

Experimental designs and results of PSO or evolutionary algorithms presented in empirical studies are sometimes based on a huge number of function evaluations ($t_{\max} > 10^5$), even for simple test functions. Our goal is to demonstrate how statistical design methods, e.g., DACE, can reduce this number significantly. The proposed approach is thoroughly analyzed for the inertia weight variant of PSO.

7.6.1 Optimizing the PSO Inertia Weight Variant

This example describes in detail how to tune the exogenous parameters of PSO. It extends the approach presented in Bartz-Beielstein et al. (2004b). Experimental designs and results presented in Shi & Eberhart (1999) have been chosen as a starting point for our analysis.

(S-1) *Preexperimental planning.* Preexperimental tests to explore the optimization potential supported the assumption that tuning might improve the algorithm's performance. RLD revealed that there exists a configuration that was able to complete the run successfully using less than 8000 function evaluations, for nearly 80% of the cases. This was less than half the number of function evaluations used in the reference study, justifying the usefulness of the analysis.

(S-2) *Scientific claim.* There exists a parameterization (design point $x^*_{\text{PSO}} \in \mathcal{D}_A$) of PSO that improves its performance significantly for one given optimization problem $x_p \in \mathcal{D}_P$.

(S-3) *Statistical hypothesis.* PSO with the parameterization x^*_{PSO} outperforms PSO with the default parameterization $x^{(0)}_{\text{PSO}}$, which is used in Shi & Eberhart (1999).

(S-4) *Specification.* Table 7.10 presents numerical results from the optimization process. Each line in Table 7.10 corresponds to one optimization step in the sequential approach. At each step, two new design points are generated and the best one is reevaluated. This is similar to the selection procedure in $(1+2)$-evolution strategies. The number of repeat runs n of the algorithm design points is increased (doubled), if a design has performed best two or more times. A starting value of $n = 2$ was chosen. For example, design point 14 performs best at iteration 1 and iteration 3. It has been evaluated 4 times, therefore the number of evaluations is set to 4 for every newly generated design. This provides a fair comparison and reduces the risk of incorrectly selecting a worse design.

Table 7.10. Problem design $x^{(1)}_{\text{rosen}}$. Inertia weight PSO optimizing the 10-dimensional Rosenbrock function. Each row represents the best algorithm design at the corresponding tuning stage. Note that function values (reported in the *second column*) can worsen (increase) although the design is improved. This happens as a result of the noise in the results, y. The probability that a seemingly good function value that is, in fact, worse might occur decreases during the sequential procedure, because the number of reevaluations is increased. The number of repeats n, is doubled if a configuration performs best twice. The corresponding configurations are marked with an asterisk

Conf	y	s	c_1	c_2	$w_{\max}$	w_{Scale}	$w_{\text{IterScale}}$	$v_{\max}$
14	6.616	26	1.457	1.989	0.713	0.482	0.684	477.874
19	18.060	39	1.302	1.843	0.871	0.273	0.831	289.922
14*	71.402	26	1.457	1.988	0.713	0.482	0.684	477.874
3	78.048	30	2.220	1.263	0.944	0.290	0.894	237.343
3*	75.615	30	2.220	1.263	0.944	0.290	0.894	237.343
35	91.094	18	1.842	1.699	0.959	0.257	0.849	95.139
43	91.544	21	1.055	1.251	0.937	0.498	0.593	681.092
52	93.754	11	1.581	2.419	0.729	0.470	0.545	98.927
20	93.997	93	1.712	1.021	0.966	0.379	0.973	11.765
19*	99.409	39	1.302	1.843	0.871	0.273	0.831	289.92
57	117.595	11	1.140	2.316	0.785	0.237	0.962	56.910
1	146.047	12	1.515	2.485	0.876	0.393	0.991	261.561
54	147.410	22	1.727	2.273	0.711	0.236	0.574	50.512
54*	98.366	22	1.727	2.273	0.711	0.236	0.574	50.512
67*	41.400	21	2.254	1.746	0.789	0.283	0.937	11.050
67*	43.225	21	2.254	1.746	0.789	0.283	0.937	11.050
67*	53.355	21	2.254	1.746	0.789	0.283	0.937	11.050

(S-5) *Experimentation.* The experimental runs are performed.

(S-6) *Statistical modeling and prediction.* Following Santner et al. (2003), the response is modeled as a realization of a regression model and a random process as described in Eq. (3.17). A Gaussian correlation function as defined in Eq. (3.18) and a regression model with polynomial of order 2 have been used. Hence, the model reads

$$Y(x) = \sum_{j=1}^{p} \beta_j f_j(x) + Z(x), \tag{7.10}$$

where $Z(\cdot)$ is a random process with mean zero and covariance $V(\omega, x) = \sigma^2 \mathcal{R}(\theta, \omega, x)$. The correlation function was chosen as

$$\mathcal{R}(\theta, \omega, z) \prod_{j=1}^{d} \exp\left(-\theta_j(\omega_j - x_j)^2\right). \tag{7.11}$$

Additionally, at certain stages a tree-based regression model as shown in Fig. 7.8 was constructed to determine parameter settings that produce outliers.

(S-7) *Evaluation and visualization.* The MSE and the predicted values can be plotted to support the numerical analysis (we produced all three-dimensional visualizations with the DACE toolbox (Lophaven et al. 2002b)). For example, the interaction between c_1 and c_2 is shown in Fig. 7.9. Values of c_1 and c_2 with $c_1 + c_2 > 4$ generated outliers that might disturb the analysis. To reduce the effects of these outliers, a design correction method has been implemented, namely $c_1 = c_2 - 4$, if $c_1 + c_2 > 4$. The right part of Fig. 7.9 illustrates the estimated MSE. Since no design point has been placed in the ranges $1 < c_1 < 1.25$ and $2.25 < c_2 < 2.5$, the MSE is relatively high. This might be an interesting region where a new design point will be placed during the next iteration. Figure 7.10 depicts the same situation as Fig. 7.9 after the determination of the design correction. In this case, a high MSE is associated with the region $c_1 + c_2 > 4$, but no design point will be placed there.

(S-8) *Optimization.* Termination or design update. Based on the expected improvement defined in Eq. (5.5), two new design points $x_{\text{PSO}}^{(1)}$ and $x_{\text{PSO}}^{(2)}$ are generated. These two designs are evaluated and their performances are compared to the performance of the current best design. The best design found so far is reevaluated. The iteration terminates if a design was evaluated for $n = 50$ times and the solution is obtained. The values in the final model read $s = 21$, $c_1 = 2.25$, $c_2 = 1.75$, $w_{\text{max}} = 0.79$, $w_{\text{scale}} = 0.28$, $w_{\text{iterScale}} = 0.94$, and $v_{\text{max}} = 11.05$. This result is shown in the last row of Table 7.10.

(S-9) *Termination.* If the obtained solution is good enough, or the maximum number of iterations has been reached, go to step (S-11).

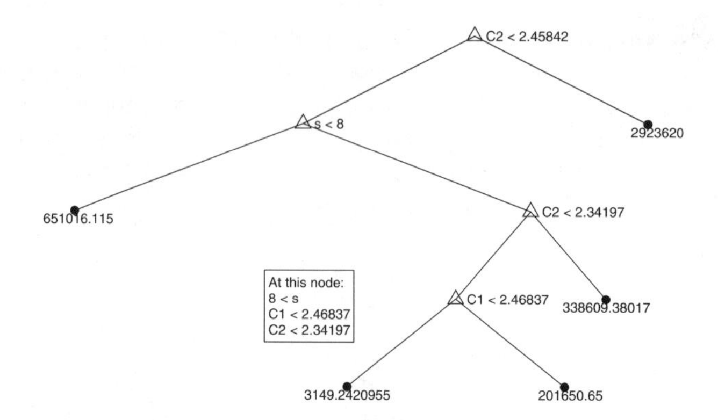

Fig. 7.8. Regression tree. Values at the nodes show the average function values for the associated node. The value in the root node is the overall mean. The left son of each node contains the configurations that fulfill the condition in the node. The configurations with $c_1 + c_2 > 4$ produce outliers that complicate the analysis. In addition, this analysis shows that the swarm size s should be larger than 8

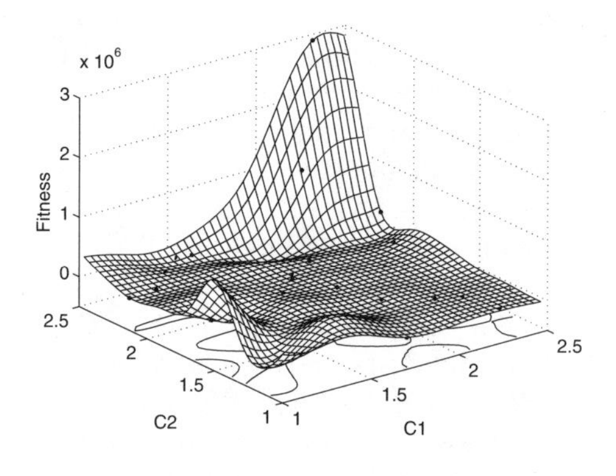

Fig. 7.9. Predicted values. Parameterizations with $c_1 + c_2 > 4$ produce outliers that complicate the analysis. The plots present results from the same data as the regression tree in Fig. 7.8

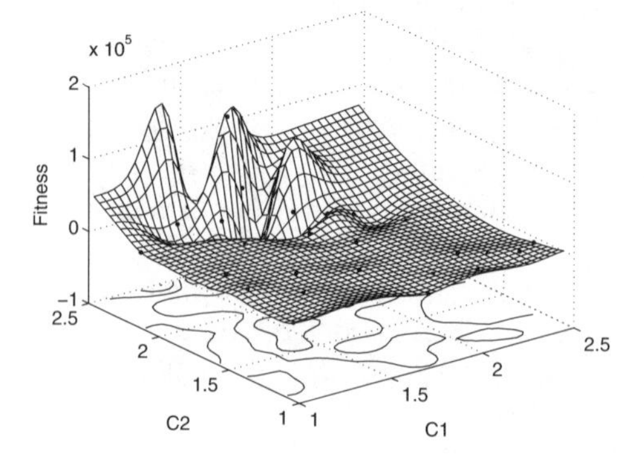

Fig. 7.10. Predicted values. The design correction avoids settings with $c_1 + c_2 > 4$ that produce outliers

(S-10) *Design update.* New design points are generated and added to the algorithm design. Go to step (S-5) to perform further experiments.

(S-11) *Rejection or acceptance.* Finally, we compare the configuration from Shi & Eberhart (1999) to the optimized configuration. The final (tuned) and the first configurations are repeated 50 times. Note, Shi & Eberhart (1999) coupled $x_{\max}$ with $v_{\max}$ as described in Eq. (6.6). The mean function value was reduced from 1.84×10^3 to 39.70, the median from 592.13 to 9.44, and the standard deviation decreased from 3.10×10^3 to 55.38. Minimum and maximum function values from 50 runs are smaller (64.64 to 0.79 and 18519 to 254.19, respectively). Histograms and boxplots are illustrated in Fig. 7.11 for both variants of PSO. The tuned design of the inertia weight PSO variant clearly improves the performance of the PSO algorithm. The statistical analysis from this and from further experiments is reported in Table 7.12. Performing a classical t-test indicates that the null hypothesis "There is no difference in the mean performances of the two algorithms" can be rejected at the 5% level.

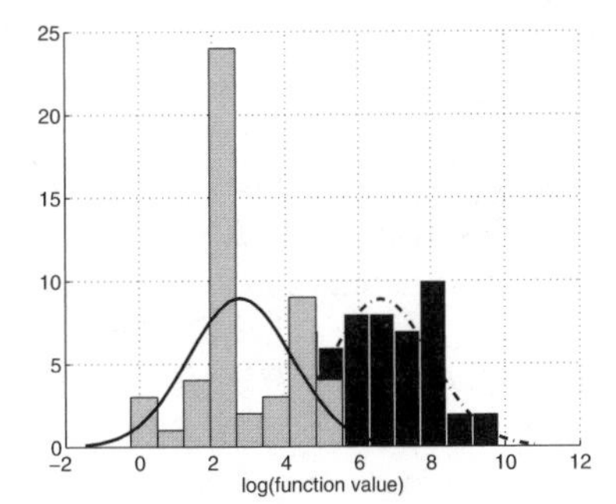

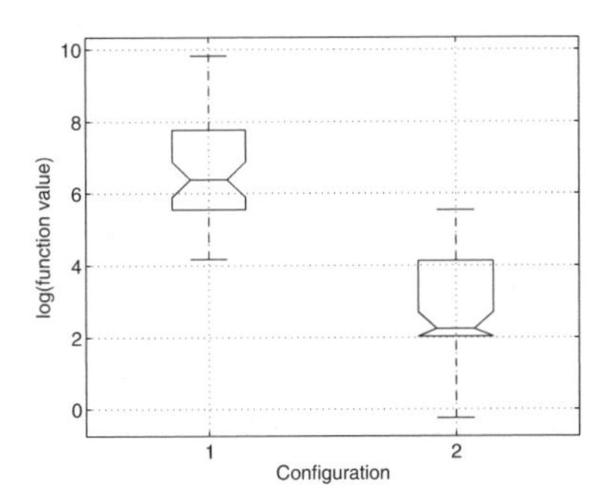

Fig. 7.11. Histogram and boxplot. *Left*: *Solid lines* and *light bars* represent the improved design. *Right*: The default configuration is denoted as 1, whereas 2 denotes the improved variant. Both plots indicate that the tuned inertia weight PSO version performs better than the default version

(S-12) *Objective interpretation.* The classical tools from exploratory data analysis such as boxplots or histograms indicate that the tuned PSO version performs better than the default PSO, cf. step (S-11). An NPT t-test comes to the same conclusion. So far we have applied methods from classical and modern statistics to test statistical hypotheses. The results indicate that the null hypothesis H should be rejected. The next step of our analysis describes how we can interpret this rejection in an objective manner and how the relationship between statistical significance and scientific import as depicted in Fig. 2.2 can be made more understandable. But before a statistical analysis is performed, we recommend looking at the raw data. Is the obtained result plausible? A comparison of the default design $x_{\text{PSO}}^{(0)}$ to the improved design x_{PSO}^{*} reveals that smaller swarm size s and $v_{\max}$ values improve the algorithm's performance for the problem design under consideration. The value of the cognitive parameter, c_1, should be increased, whereas the value of the social parameter, c_2, should be reduced. The parameters related to the scaling of the inertia weight, w, should be reduced, too. The improved design does not contain any exceptional parameter settings. It appears to be a reasonable choice. Methods that answer the question "*Why* does an increased value of c_1 lead to a better performance?" will be discussed in Chap. 8. Here, we are interested in the question "Does algorithm A outperform algorithm B?" But how can we be sure that the related PSO performance is better than the performance of the default PSO? Error statistical tools can be used to tackle this problem. How do NPT* interpretations go beyond the results found with NPT tools? This question is closely related to the problem stated in Example 2.1 in the first part of this book.

We claim that statistical tests are means of learning. We are interested in detecting differences between the correct model and a hypothesized one. Experiments provide means to observe the difference between a sample statistic and a hypothesized population parameter. The distribution of the test statistic S can be used to control error probabilities. A rejection of a statistical hypothesis H can be misconstrued if it is erroneously taken as an indication that a difference of scientific importance has been detected. Plots of the observed significance as introduced in Sect. 2.5 are valuable tools that can be used to detect whether this misconstrual occurs and to evaluate the scientific meaning.

We will consider the case of rejecting a hypothesis H first. An NPT* interpretation of accepting a hypothesis will be discussed in Sect. 7.6.2.

The relationship between the observed significance level $\alpha_{\overline{d}}(\delta)$, the difference in means $\delta = \mu_1 - \mu_2$ of the default PSO, and the tuned PSO version is illustrated in Fig. 7.12. First consider the values of the observed significance level ($\alpha_{\overline{d}}(\delta)$) for a sample size of $n = 50$, where $\overline{d}$ denotes the observed difference in means and δ the hypothesized difference as introduced in Sect. 2.5. The observed difference in means is $\overline{d} = 1798.6$; one standard deviation unit has the value $\sigma_{\hat{x}} = 410.84$. A t-test indicates that the null hypothesis can be rejected at the 5% level. Does the observed difference in means occur due to the experimental error only and is the rejection of the null hypothesis H misconstrued? A difference in means of less than 1 or 2 standard error units might be caused by experimental errors. Observe the values of $\alpha(1798.6, \delta) = \alpha_{1798.6}(\delta)$. How often does a rejection arise when various populations are observed? Since an observed significance level $\alpha_{1798.6}(410.84) = 0.01$ is small, it is a good indication that we are observing two populations with a difference in means $\delta > 410.84$. If one observes a difference $\overline{d}$ when the true difference in means δ was no greater than 410.84, only 1% of the observed differences would be this large. This gives good reason to reject the associated null hypothesis

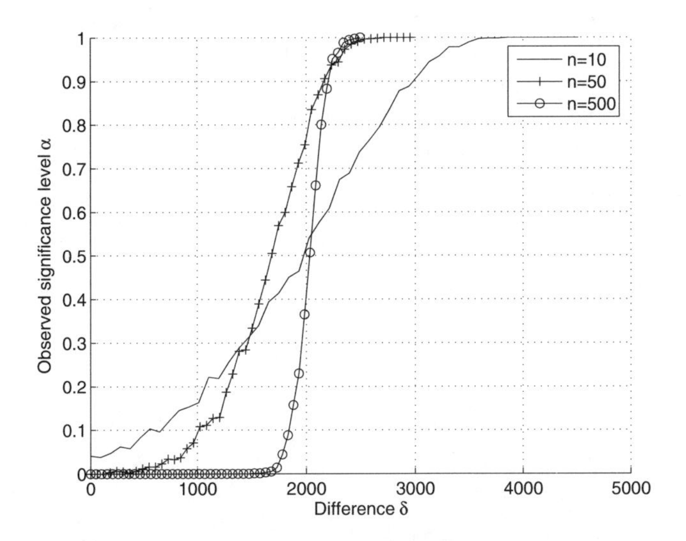

Fig. 7.12. Observed significance level. The particle swarm optimization with the default parameters is compared to the tuned version. The observed differences are 1933, 1799, and 2077 for $n = 10$, 50, and 500 experiments, respectively (function value while optimizing the 10-dimensional Rosenbrock function with a budget of 2500 function evaluations). Consider $n = 50$: A difference as small as 1100, which would occur frequently, has an observed significance level smaller than 0.05. This is a strong indication for the assumption that there is a difference as large as 1000. This is case RE-2.1 as defined in Sect. 2.5

$H : \delta \leq 410.84$. And, we can learn even more from this result: It also is an indication that $\delta > 822$, since $\alpha_{1798.6}(828) \approx 0.05$. The situation depicted in Fig. 7.12 is similar to the situation discussed in Example 2.5. Low $\alpha_{\overline{d}}(\delta)$ values are not due to large sample sizes only. Therefore the statistical results indicate that there is a difference in means and this difference is also scientifically meaningful.

7.6.2 Optimizing the PSO Constriction Factor Variant

The design of the PSO constriction factor variant was tuned in a similar manner as the inertia weight variant. The initial LHD is reported in Table 7.11, where $x^{(l)}_{\text{PSOC}}$ and $x^{(u)}_{\text{PSOC}}$ denote the lower and upper bounds of the experimental region, respectively, x^*_{PSOC} is the improved design that was found by the sequential procedure, and $x^{(0)}_{\text{PSOC}}$ is the default design recommended in Clerc & Kennedy (2002). The run length distributions shown in Fig. 7.13 do not clearly indicate which configuration performs better. Although the curve of the tuned version (constriction*) is above the curve of the default variant (constriction), it is not obvious whether this difference is significant.

Table 7.11. PSO constriction factor. Algorithm designs to optimize Rosenbrock's function. The variables s, χ, φ, and $v_{\max}$ have been defined in Table 6.5. $x_{\text{PSOC}}^{(l)}$ and $x_{\text{PSOC}}^{(u)}$ denote the ranges of the LHD, x_{PSOC}^{*} is the improved design, and $x_{\text{PSOC}}^{(0)}$ is the design suggested in Clerc & Kennedy (2002)

Design	s	χ	φ	$v_{\max}$
$x_{\text{PSOC}}^{(l)}$	5	0.68	3.0	10
$x_{\text{PSOC}}^{(u)}$	100	0.8	4.5	750
x_{PSOC}^{*}	17	0.759	3.205	324.438
$x_{\text{PSOC}}^{(0)}$	20	0.729	4.1	100

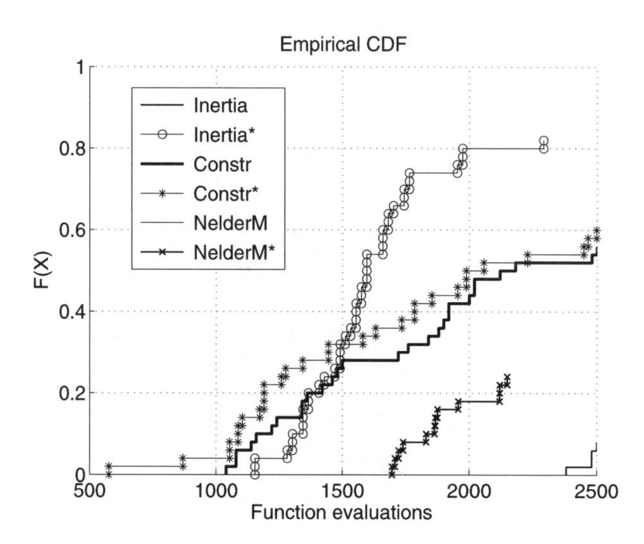

Fig. 7.13. Run length distribution. Step (S-11), the final comparison of the canonical and the improved design based on RLDs. *Asterisks* denote improved configurations. The improved inertia weight version of PSO succeeded in more than 80% of the experiments with less than 2500 function evaluations. The standard NMS algorithm failed completely (hence the corresponding curve is not shown in this figure), but it was able with an improved design to succeed in 10% of the runs after 2500 function evaluations. For a given budget of 1500 function evaluations, both the constriction factor and the improved inertia weight PSO variants perform equally well

The numerical values indicate that the tuned version performs slightly better than the default one (106.56 versus 162.02, as can be seen in Table 7.12), but the corresponding graphical representations (histograms and boxplots, not shown here) give no hints that there is a significant difference between the performance of the tuned x_{PSOC}^{*} and $x_{\text{PSOC}}^{(0)}$ (Clerc & Kennedy 2002). This result is not very convincing. Further investigations are recommended.

However, a t-test would accept the null hypothesis that there is no difference in means. But, is this result independent of the sample size? If the sample size is increased, for example, if 2000 experiments were performed, a t-test would reject the null hypothesis at the 5% level. This example demonstrates how the experimenter can influence the outcome of the classical t-test by varying the sample size n. Figure 7.14 illustrates the situation with tools from the arsenal of an error statistician. The result presented in Fig. 7.14 is

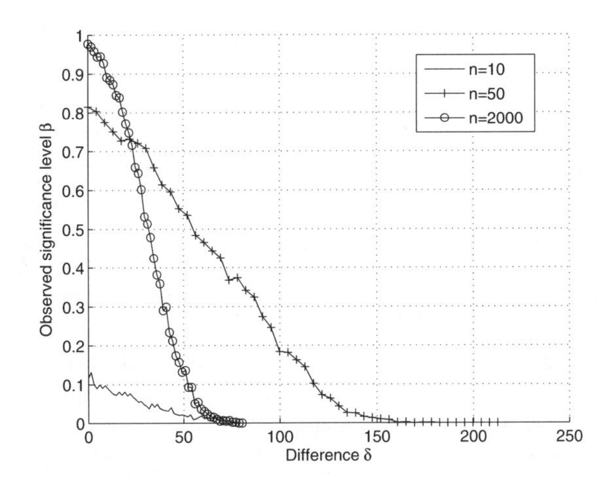

Fig. 7.14. Comparing PSO constriction and PSO constriction*, $n = 10$, 50, and 2000 repeats. The observed difference for $n = 50$ is $\overline{d} = 55.47$. A t-test would accept the null hypothesis H "there is no difference in means" for $n = 50$, because $\beta_{55.47}(0) < 0.95$. A t-test would reject the null hypothesis for $n = 2000$, because $\beta_{30.1}(0) > 0.95$. This is case AC-2.1 as introduced in Sect. 2.5. The t-test results depend on the sample size

Table 7.12. Result table of the mean function values of the best particle in the swarm after $n = 50$ runs, $\tilde{f}^{(50)}$, for the Rosenbrock function. Q-N is the quasi-Newton methods from Sect. 6.1.2. Default algorithm designs from Shi & Eberhart (1999); Clerc & Kennedy (2002); Lagarias et al. (1998), as well as the improved design for all algorithms, for $n = 50$ runs, are reported

Design	Mean	Median	SD	Min	Max
$x^{(0)}_{\text{PSO}}$	1.84×10^3	592.13	3.10×10^3	64.64	18519
x^{*}_{PSO}	**39.70**	**9.44**	**55.38**	**0.79**	**254.19**
$x^{(0)}_{\text{PSOC}}$	162.02	58.51	378.08	4.55	2.62×10^3
x^{*}_{PSOC}	106.56	37.65	165.90	0.83	647.91
$x^{(0)}_{\text{NMS}}$	9.07×10^3	1.14×10^3	2.50×10^4	153.05	154966
x^{*}_{NMS}	112.92	109.26	22.13	79.79	173.04
Q-N	5.46×10^{-11}	5.79×10^{-11}	8.62×10^{-12}	1.62×10^{-11}	6.20×10^{-11}

a good indication that we are observing a population where the difference in means is not larger than 120 ($n = 50$), or not larger than 50 ($n = 2000$).

But is this result really scientifically important? If the experimenter has specified the largest scientifically unimportant difference greater than zero, then this can be used to relate the statistical result to the scientific claim. Obviously, metastatistical rules are necessary to interpret this result.

7.6.3 Comparing Particle Swarm Variants

Our next goal is to detect differences between the two major particle swarm variants, the inertia weight and the constriction factor variant. As the former requires only four parameters, a legitimate question is "Why does the inertia weight variant require three additional factors?" We consider the differences in the performance of the constriction and inertia weight particle swarm optimization variants based on optimization data from the 10-dimensional Rosenbrock function. Fifty experiments were performed, resulting in 50 differences. The null hypothesis reads: "There is no difference in means." The observed difference is 66.86. As histograms and boxplots reveal, there is no statistically significant difference observable. Both configurations perform similarly. Population mean, median, and standard deviation are shown in Table 7.12.

The plot of the observed significance versus the difference in means indicates that differences δ in the mean value $\tilde{f}^{(50)}$ larger than 100 seldom occur. If the experimenter specifies the smallest scientifically significant difference he can judge the consequences of accepting the null hypothesis.

7.6.4 Optimizing the Nelder–Mead Simplex Algorithm and a Quasi-Newton Method

In the Nelder–Mead simplex algorithm, four parameters must be specified, namely the coefficients of reflection ρ, expansion χ, contraction γ, and shrinkage σ. Default settings are reported in Table 6.1. Experiments indicate that the value of the reflection parameter, ρ, should be smaller than 1.5. An analysis that is based on the visualization tools from the DACE toolbox reveals that there exists a relatively small local optimum regarding χ (expansion parameter) and ψ (contraction parameter), respectively. The sequential approach could be successfully applied to the NMS algorithm, its performance on the Rosenbrock function was improved significantly. Results from this tuning process are presented in Table 7.12.

In addition to the optimization algorithms analyzed so far, the performance of a quasi-Newton method (see Sect. 6.1.2) was analyzed. An implementation from the commercial MATLAB optimization toolbox was used in the experiments. Quasi-Newton clearly outperformed the other algorithms, as can be seen from the results in Table 7.12.

A comparison of the RLDs of the three algorithms is shown in Fig. 7.13. The results support the claim that PSO performs better than the NMS algorithm. Only the tuned version of the latter was able to complete 20% of the experiments with success. Regarding the two PSO variants, it is not obvious which one performs better. After the tuning process, the inertia weight variant appears to be better, but it requires the specification of seven (compared to only four in the constriction factor variant) exogenous parameters. However, the Rosenbrock function is mostly of academic interest, since it lacks many features of a real-world optimization problem.

The analysis and the tuning procedure described so far have been based solely on the average function value in 50 runs. This value may be irrelevant in a different optimization context. For example, the best function value (minimum) or the median can be alternatively used. A similar optimization procedure could have been performed for any of these cases with the presented sequential approach. Note that the optimal algorithm design presented in this study is only applicable to this specific optimization task $x^{(1)}_{\texttt{rosen}}$ as listed in Table 7.8.

As in Shi & Eberhart (1999), the starting points have been initialized randomly in the range $[15, 30]^d$. Hence, different sources of randomness are mixed in this example. The following studies will be based on deterministically generated starting points, as recommended in More et al. (1981).

7.7 Experimental Results for the S-Ring Model

The Rosenbrock function, which was considered in the previous sections, was chosen to provide a comprehensive introduction to the sequential DACE approach. In the following, we will present a more realistic real-world problem. The performance of a PSO is compared to a NMS algorithm and to a quasi-Newton method. The S-ring simulator was selected to define a 12-dimensional optimization problem with noisy function values. The number of function evaluations, $t_{\max}$, was limited to 1000 for each optimization run. This value appears to be realistic for real-world applications. The related problem design is reported in Table 7.13.

Similar to the analysis for the Rosenbrock function, the constriction factor and inertia weight variants of PSO were analyzed. The former requires only four exogenous strategy parameters, whereas seven parameters have to be

Table 7.13. Problem design for the S-ring experiments. Note that due to the stochastic nature of the S-ring simulation model, no additional noise was added to the function values

Design	Init.	Term.	PM	n	$t_{\max}$	d	x_l	x_u
$x^{(1)}_{\texttt{sring}}$	DETMOD	EXH	MBST	50	1000	12	−10	10

specified for the latter. Optimizing the PSO inertia weight variant improved the algorithm's robustness as reported in Table 7.14. The average function value decreased from 2.61 to 2.51, which is a significant difference. However, it is very important to note that the minimum function value could not be improved, but increased slightly from 2.4083 to 2.4127. The tuning procedure was able to find an algorithm design that prevents outliers and produces robust solutions at the cost of an aggressive exploratory behavior. However, an increased probability of finding a solution that has a minimum function value could have been specified as an optimization goal, resulting in a different "optimal" design. Measures such as the best solution from n runs are better suited for real-world optimization problems than the mean function value. Computer-intensive methods facilitate the determination of related statistics.

Although the function values look slightly better, the tuning process produced no significant improvement for the rest of the algorithms. The constriction factor PSO variant, as well as the NMS algorithm and the quasi-Newton method were not able to escape from local optima. In contrast to the Rosenbrock function, many real-world optimization problems have many local minima on flat plateaus. The distribution of local optima in the search space is unstructured. Therefore these algorithms were unable to escape plateaus of equal fitness. This behavior occurred independently from the parameterization of their exogenous strategy parameters. The inertia weight PSO variant that required the determination of seven exogenous strategy parameters outperformed the other algorithms in this comparison. Whether this improved result was caused by the scaling property of the inertia weight is subject to further investigation.

Experimental results indicate that there is no generic algorithm that works equally well on each problem. Even different instances of one problem may

Table 7.14. Results from the optimization of the S-ring model. Default designs, reported in Shi & Eberhart (1999); Clerc & Kennedy (2002); Lagarias et al. (1998), and improved designs, for $n = 50$ repeats, are reported. The tuned inertia weight PSO variant appears to be more robust than the default variant. It generates no outliers, as can be seen in the *last column*, but it was not able to find a better minimum. This is understandable, because the sequential parameter optimization selected algorithm designs based on their mean best function values

Design	Mean	Median	SD	Min	Max
$x_{\text{PSO}}^{(0)}$	2.6152	2.5726	0.4946	**2.4083**	5.9988
x_{PSO}^{*}	**2.5171**	**2.5112**	**0.0754**	2.4127	**2.6454**
$x_{\text{PSOC}}^{(0)}$	4.1743	2.6252	1.7021	2.5130	5.9999
x_{PSOC}^{*}	4.1707	2.6253	1.7055	2.5164	5.9999
$x_{\text{NMS}}^{(0)}$	4.3112	4.3126	1.7059	2.6200	5.9999
Quasi-Newton	4.3110	4.3126	1.7060	2.6186	5.9999

require different algorithms, or at least different parameterizations of the employed algorithms. None of the algorithms has proved in our study to be satisfying for every problem. The quasi-Newton method, as expected, outperformed the other algorithms on the Rosenbrock function, but it failed completely on the elevator optimization problem, where the inertia weight PSO variant, which requires nearly twice as many parameters as the PSO constriction factor variant, performed best.

Finally, we note that the determination of a good initial algorithm (and problem) design is not trivial, and therefore is a drawback of the proposed approach. This is common to all statistical methods in this field, especially for the classical DOE approach.

7.8 Criteria for Comparing Algorithms

Nowadays it is widely accepted that there is no algorithm that performs on average better than any other algorithm. Schwefel (1995) comments on evolution strategies:

> So, is the evolution strategy the long-sought-after *universal* method of optimization? Unfortunately, things are not so simple and this question cannot be answered with a clear "yes."

Some optimization algorithms are exceptionally popular, for example, the Nelder–Mead simplex algorithm or evolutionary algorithms. The popularity of these direct search algorithms is not founded on their overall optimality, but might be related to the following reasons:

1. Direct search algorithms are appealing, because they are easy to explain, understand, and implement. They share this feature with some of the designs presented in Chap. 5.
2. For many real-world optimization problems, it is vital to find an improvement, but not the global optimum. Direct search algorithms produce significant improvements during the first stage of their search.
3. Function evaluations are extremely costly in many real-world applications. Hence, the usage of finite-gradient approximation schemes that require at least d function evaluations in every step is prohibitive (d denotes the problem dimension).

We claim that universal optimization methods are suitable tools during the first stage of an optimization process. The experimental methodology presented in this chapter provides statistical tools to detect relevant factors. It can be advantageous to combine or even to replace the universal method with small, smart, and flexible heuristics. The experimental analysis provides means for a deepened understanding of the problem, the algorithm, and their interaction as well. Learning happens and leads to progress in science.

Table 7.15. Sequential parameter optimization. This approach combines methods from computational statistics and exploratory data analysis to improve (tune) the performance of direct search algorithms. It can be seen as an extension of the guidelines from experimental algorithmics presented in Chap. 1

Step	Action
(S-1)	Preexperimental planning
(S-2)	Scientific claim
(S-3)	Statistical hypothesis
(S-4)	Specification of the
	(a) Optimization problem
	(b) Constraints
	(c) Initialization method
	(d) Termination method
	(e) Algorithm (important factors)
	(f) Initial experimental design
	(g) Performance measure
(S-5)	Experimentation
(S-6)	Statistical modeling of data and prediction
(S-7)	Evaluation and visualization
(S-8)	Optimization
(S-9)	Termination: If the obtained solution is good enough, or the maximum number of iterations has been reached, go to step (S-11)
(S-10)	Design update and go to step (S-5)
(S-11)	Rejection/acceptance of the statistical hypothesis
(S-12)	Objective interpretation of the results from step (S-11)

7.9 Summary

The ideas presented in this chapter can be summarized as follows:

1. Tuning was introduced as an optimization process.
2. Optimization relies on very restrictive assumptions. "With the possible exception of the laboratory or casino" these assumptions are met nowhere (Klein 2002).
3. An optimization process can be regarded as a process that enables learning. This concept is related to Mayo's extension of the classical NPT approach.

4. To start the tuning process, a performance measure has to be defined. Effectivity and efficiency can guide the choice of an adequate performance measure.
5. The classical DOE approach consists of three steps: screening, modeling, and optimization. Each step requires different experimental designs.
6. As the assumption of a linear model for the analysis of computer programs is highly speculative, a sequential approach (SPO) that combines classical and modern statistical tools has been proposed. This sequential process can be used for tuning algorithms. It consists of the twelve steps that are reported in Table 7.15.
7. Results that are statistically significant are not automatically scientifically meaningful. Results from the classical Neyman–Pearson theory of testing should be complemented with NPT* tools.
8. The optimization practitioner does not always choose the absolute best algorithm. Sometimes a robust algorithm or an algorithm that provides insight into the structure of the optimization problem is preferred.

7.10 Further Reading

Montgomery (2001) discusses classical DOE techniques, whereas Santner et al. (2003) give an introduction to DACE. Breiman et al. (1984) is a standard text book for classification and regression tree methods. Klein's viewpoint is based on the approach of bounded rationality (Simon 1955; Rubinstein 1998; Klein 2002).

The tuning approach presented in this chapter has been successfully applied to several optimization tasks, for example, in evolutionary optimization of mold temperature control strategies (Mehnen et al. 2004a), digital circuit design using evolutionary algorithms (Beielstein et al. 2001, 2002a), elevator group control (Beielstein et al. 2003a; Bartz-Beielstein et al. 2005c), genetic programming of algorithmic chemistry solving the 4-bit odd parity problem (Lasarczyk & Banzhaf 2005a, b), or real-world problem from the chemical engineering domain, the design of a nonsharp separation sequence (Aggarwal & Floudas 1990; Bartz-Beielstein et al. 2005b)

Bartz-Beielstein (2003) uses the classical DOE approach to compare a variant of simulated annealing (Belisle 1992) to an evolution strategy. An approach that combines classical DOE techniques, regression trees, and DACE was shown in Bartz-Beielstein & Markon (2004).

Bartz-Beielstein et al. (2003a), Mehnen et al. (2004a, b), Weinert et al. (2004), Bartz-Beielstein & Naujoks (2004), and Bartz-Beielstein et al. (2004c) applied tuning procedures to multicriteria optimization problems.

8

Understanding Performance

> Life is really simple, but men insist on making it complicated.
> —Confucius

This chapter closes the circle on the problem begun in the discussion of the Neyman–Pearson theory in Chap. 2. It demonstrates the difference between statistical testing as an automatic rule (NPT) and as a learning tool (NPT*). Automatic rules can be implemented as computer programs that generate solutions. Learning tools provide means to interpret the relevance of these results.

First, existing approaches and theoretical results for the design and analysis of experiments for selection and screening are presented. The related statistical procedures require assumptions that are not always met in practice, especially when applied to search algorithms. A classification of methods that can be integrated into the selection process of evolutionary algorithms is presented. Threshold selection is one approach to handle the problem of noisy function evaluations. The second part presents a case study to demonstrate how NPT* tools enable an understanding of the basic principles of threshold selection.

Understanding can be seen not only as an analytic, but also as a synthetic, bottom-up approach: Simple algorithms often perform excellently on realistic benchmark problems (Watson et al. 1999; Whitley et al. 2002). Therefore, it might be useful to determine the essential features of algorithms. Starting from the very basic parts of the algorithm, new parts are added if they improve the algorithm's performance. Interactions play an important role, since some effects may appear only as a result of correlations between two or more parts. In preexperimental studies, simple configurations are tested to find out whether there is any effect at all. They enable the experimenter to define a first experimental design and to state the scientific claim more precisely.

8.1 Selection Under Uncertainty

Noise is a common factor in most real-world optimization problems. It arises from different sources, such as measurement errors in experiments, the stochastic nature of the simulation process, or the limited number of samples gathered from a large search space. Common means used by evolutionary algorithms to cope with noise are resampling, averaging techniques based on statistical tests, local regression methods for function value estimation, or methods to vary the population size (Stagge 1998; Beyer 2000; Sano & Kita 2000; Arnold 2001; Branke et al. 2001; Bartz-Beielstein & Markon 2004). In this book we concentrate our investigations on the selection process when the function values are disturbed by additive noise.

Noise that affects the object variables is not the subject of our investigations. From our point of view the following case is fundamental for the selection procedure in noisy environments (Markon et al. 2001):

> Reject or accept a new candidate, while the available information is uncertain. Thus, two errors may occur: An α error as the probability of accepting a worse candidate due to noise and a β error, as the error probability of rejecting a better candidate.

In the context of selection and decision making, the terms "candidate" and "point" will be used synonymously. A well-established context where these error probabilities are analyzed is hypothesis testing as introduced in Sect. 3.1.

8.1.1 A Survey of Different Selection Schemes

Depending on the prior knowledge, selection schemes can be classified according to the following criteria:

1. threshold: subset selection—indifference zone
2. termination: single stage—multistage (sequential)
3. sample size: open procedures—closed procedures
4. variances: known—unknown, equal—unequal

The goal of *subset selection* is the identification of a subset containing the best candidate. It is related to screening procedures. Subset selection is used when analyzing results, whereas the *indifference zone* approach is used when designing experiments. The sample size r is known in subset selection approaches; it is determined prior to the experiments in the indifference zone approaches.

Single-stage procedures can be distinguished from *multistage* procedures. The terms "multistage" and "sequential" will be used synonymously. The latter can use *elimination*: If inferior solutions are detected, they are eliminated immediately. Selection procedures are *closed* if prior to experimentation an upper bound is placed on the number of observations to be taken from each candidate. Otherwise, they are *open*. Furthermore, it is important to know whether the variance is common or known.

Our analysis is based on the following statistical assumptions. Let $\{Y_{ij}\}$, $1 \leq i \leq r$, $1 \leq j \leq s$, denote r independent random samples of observations, taken from $s \geq 2$ candidates. The Y_{ij} can denote function values taken from candidate solutions $X_1, \ldots, X_s$ or individuals (particles) of some evolutionary algorithm. Candidate X_i has a (fitness) function value with unknown mean μ_i and common unknown variance $\sigma^2_{\epsilon,i} = \sigma^2_\epsilon$, $1 \leq i \leq s$. The *ordered means* are denoted by

$$\mu_{[1]} \leq \mu_{[2]} \leq \ldots \leq \mu_{[s]}, \tag{8.1}$$

where $\mu_{[1]}$ denotes the mean of the best candidate (minimization). Generally, normal response experiments are considered.

8.1.2 Indifference Zone Approaches—A Single Stage Procedure

In the indifference zone approach, the optimization practitioner a priori specifies a value $\delta^* > 0$ representing the smallest difference worth detecting (threshold). The difference δ^* is related the largest scientifically unimportant value δ_{un} (Eq. 2.9). Errors below this threshold resulting from incorrect selection are ignored. Following Bechhofer et al. (1995) we define *experimental goals* (G) and associated *probability requirements* (P). The experimental goal is related to the scientific claim (see step (S-2) in Sect. 7.5), whereas the probability requirement is related to the statistical model of experimental tests (see Fig. 2.2). The first experimental goal reads:

(G-8.1) To select the candidate associated with the smallest mean $\mu_{[1]}$.

A *correct selection* (CS) is said to have been made if (G-8.1) is achieved. Let δ^*, $0 < \delta^* < \infty$, be the smallest difference worth detecting. The *probability requirement* reads

(P-8.1) For given constants (δ^*, P^*) with $1/s < P^* < 1$, we require

$$\Pr(\text{CS}) \geq P^*, \qquad \text{whenever } \mu_{[2]} - \mu_{[1]} \geq \delta^*. \tag{8.2}$$

A configuration that satisfies the *preference zone requirement*

$$\mu_{[2]} - \mu_{[1]} \geq \delta^*, \tag{8.3}$$

is said to be in the *preference zone*, otherwise it is said to be in the *indifference zone*. Indifference zone approaches are procedures that guarantee Eq. (8.2). Bechhofer et al. (1995) proposed the single-stage selection procedure shown in Fig. 8.1 for common known variance. The selection procedure shown in Fig. 8.1 is location-invariant; only the difference in means, and not their absolute values are important. The upper-α equicoordinate critical point $Z^{(\alpha)}_{s,\rho}$, see Eq. (3.4), is determined to satisfy the probability requirement (P-8.1) for any true configuration of means satisfying the preference zone requirement, see Eq. (8.3). Under the assumptions from Sect. 8.1.1 is no procedure requiring fewer observations per candidate than the procedure shown in Fig. 8.1 if the experimenter is restricted to single-stage location-invariant procedures that guarantee the probability requirement (P-8.1) (Bechhofer et al. 1995).

Procedure: Single-stage procedure

1. For the given s and specified $(\delta^*/\sigma_\epsilon, P^*)$ determine

$$r = \left\lceil 2 \left(\sigma_\epsilon Z^{(1-P^*)}_{s-1,1/2} / \delta^* \right)^2 \right\rceil . \tag{8.4}$$

2. Take a random sample of r observations Y_{ij}, $1 \leq j \leq r$, in a single stage from X_i, $1 \leq i \leq s$.
3. Calculate the s sample means $\overline{y}_i = \sum_{j=1}^{r} y_{ij}/r$, $1 \leq i \leq s$.
4. Select the candidate that yields the smallest sample mean $\overline{y}_{[1]}$ as the one associated with the smallest sample mean $\mu_{[1]}$.

Fig. 8.1. Indifference zone approach; single-stage procedure

8.1.3 Subset Selection

Selection of a Subset of Good Candidates

A common problem for population-based direct search methods is the selection of a subset of s "good" candidates out of a set of m $(1 \leq s < m)$ under uncertainty. Gupta (1965) proposed a single-stage procedure, which is applicable when the function values of the candidates are balanced (see also Sect. 3.1.2) and normal with common variance.

Selection of a Random-Size Subset of Good Candidates

This selection method generates a random-size subset that contains the candidate associated with the smallest true mean $\mu_{[1]}$.

(G-8.2) To select a (random-size) subset that contains the candidate X_i associated with $\mu_{[1]}$.

For unknown variance σ_ϵ^2 the probability of a correct selection depends on (μ, σ_ϵ^2). If the variance is known, Pr(CS) depends only on the true means $\mu = (\mu_1, \ldots, \mu_s)$.

(P-8.2) For a specified constant P^* with $1/s < P^* < 1$, we require that

$$\Pr\{\text{CS}|(\mu, \sigma_\epsilon^2)\} \geq P^* \tag{8.5}$$

for all μ.

The Gupta selection procedure shown in Fig. 8.2 implements a (random-size) subset-selection method. Bartz-Beielstein & Markon (2004) implemented this selection scheme for evolutionary algorithms in noisy environments. As the size of the selected subset is not known in advance, the population size varies during the optimization: It increases with the noise level. Nelson et al. (1998) and Goldsman & Nelson (1998) propose an extension of Gupta's single-stage procedure that is also applicable if the variances are unknown and not necessarily equal. A subset-selection approach for the selection of the s best candidates is described in Bechhofer et al. (1995, p. 86).

Procedure: Gupta selection for unknown variance

1. Take a random sample of r observations Y_{ij}, $1 \le j \le r$, in a single stage from X_i, $1 \le i \le s$.
2. Calculate the s sample means $\overline{y}_i = \sum_{j=1}^{r} y_{ij}/r$, $1 \le i \le s$.
3. Calculate

$$s_\nu^2 = \sum_{i=1}^{s}\sum_{j=1}^{r}(y_{ij} - \overline{y}_i)^2/\nu, \tag{8.6}$$

   the unbiased pooled estimate of σ_ϵ^2 based on $\nu = s(r-1)$ degrees of freedom.
4. Include the candidate X_i in the selected subset if and only if

$$\overline{y}_i \le \overline{y}_{[1]} + h s_\nu \sqrt{2/r}, \tag{8.7}$$

   where

$$h = T_{s-1,\nu,1/2}^{(1-P^*)}. \tag{8.8}$$

Fig. 8.2. Subset selection. This is a single-stage procedure for unknown variance σ_ϵ^2; h is the equicoordinate critical point of the equicorrelated multivariate t-distribution. If σ_ϵ^2 is known, h is the upper $(1 - P^*)$ equicoordinate critical point of the equicorrelated multivariate standard normal distribution, see Eq. (3.4)

Selection of δ^*-Near-Best Candidates

Selecting the near-best candidate may be more useful than selecting the s best candidates in some situations (Fabian 1962; Bechhofer et al. 1995). Candidate X_i is δ^*-near-best, if μ_i is within a specified amount $\delta^* > 0$ of the smallest sample mean:

$$\mu_i \le \mu_{[1]} + \delta^*. \tag{8.9}$$

(G-8.3) Select a (random-size) subset that contains at least one candidate X_i satisfying Eq. (8.9).

(P-8.3) For specified constants (δ^*, P^*) with $\delta^* > 0$ and $1/s < P^* < 1$, we require that (see Roth (1978); van der Laan (1992)):

$$\Pr\{\delta^*\text{-near-best CS}\} \ge P^*, \tag{8.10}$$

for all μ.

A δ^*-near-best selection procedure is shown in Fig. 8.3.

Procedure: δ^*-Near-best selection

1. Take a random sample of r observations Y_{ij}, $1 \leq j \leq r$, in a single stage from X_i, $1 \leq i \leq s$.
2. Calculate the s sample means $\overline{y}_i = \sum_{j=1}^{r} y_{ij}/r$, $1 \leq i \leq s$.
3. Include the candidate X_i in the selected subset if and only if

$$\overline{y}_i \leq \overline{y}_{[1]} + h(\delta^*)\sigma_\epsilon\sqrt{2/r}, \tag{8.11}$$

where

$$h(\delta^*) = Z_{s-1,1/2}^{(1-P^*)} - \delta^*/\sigma_\epsilon\sqrt{r/2}. \tag{8.12}$$

Fig. 8.3. δ^*-Near-best selection

8.1.4 Threshold Selection

Threshold rejection (TR) and *threshold acceptance* (TA) are complementary strategies. Threshold rejection is a selection method for evolutionary algorithms that accepts new candidates if their noisy function values are significantly better than the value of the other candidates (Markon et al. 2001). "Significant" is equivalent to "by at least a margin of τ." The threshold value τ is related the largest scientifically unimportant value δ_{un} (Eq. 2.9). Threshold acceptance accepts a new candidate even if its noisy function value is worse. The term "threshold selection" subsumes both selection strategies.

The basic idea of threshold selection is relatively simple and is already known in other contexts:

- Matyáš (1965) introduced a threshold operator (with some errors, see Driml & Hanš (1967)) for a (1+1)-evolution strategy and objective functions without noise.
- Stewart et al. (1967) proposed a threshold strategy that accepts only random changes that result in a specified minimum improvement in the function value.
- Dueck & Scheuer (1990) presented a threshold acceptance algorithm.
- Winker (2001) discussed threshold acceptance for problems in econometrics, statistics and operations research.
- Nagylaki (1992) stated that a similar principle, the so-called truncation selection, is very important in plant and animal breeding: "Only individuals with phenotypic value at least as great as some number c are permitted to reproduce." Truncation selection is important for breeders, but it is unlikely to occur in natural populations.

Threshold selection is also related to Fredkin's paradox: "The more equally attractive two alternatives seem, the harder it can be to choose between them—no matter that, to the same degree, the choice can only matter less" (Minsky 1985). Regarding the distinction between rules of inductive behavior and learning rules given in Sect. 2.5.2, TS as presented here is an automatic test rule and belongs to the former type of rules.

The Threshold Selection Procedure

As in (G-8.1), the experimental goal is to select the candidate associated with the smallest mean $\mu_{[1]}$. Figure 8.4 shows the threshold selection algorithm. As can be seen from Eq. (8.13), threshold rejection increases the chance of

Procedure: Threshold selection

1. Given: A candidate X_1 with a related sample Y_{1j} of r observations and sample mean $\overline{y}_1 = \sum_{j=1}^{r} y_{1j}/r$.
2. Take a random sample of r observations Y_{2j}, $1 \leq j \leq r$, in a single stage from a new candidate X_2.
3. Calculate the sample mean $\overline{y}_2 = \sum_{j=1}^{r} y_{2j}/r$.
4. Select the new candidate X_2 if and only if

$$TR: \overline{y}_2 + \tau < \overline{y}_1, \qquad \text{with } \tau \geq 0, \tag{8.13}$$

or

$$TA: \overline{y}_2 + \tau < \overline{y}_1, \qquad \text{with } \tau \leq 0. \tag{8.14}$$

Fig. 8.4. Threshold selection. This basic procedure can be implemented in many optimization algorithms, for example, evolution strategies or particle swarm optimization. TA increases the chance of accepting worse candidates, whereas TR accepts solutions if their observed function value is better by at least a margin τ than the best value observed so far

rejecting a worse candidate at the expense of accepting a good candidate. It might be adequate if there is a very small probability of generating a good candidate. Equation (8.14) reveals that threshold acceptance increases the chance of accepting a good candidate at the risk of failing to reject worse candidates.

Threshold Selection and Hypothesis Testing

The calculation of a threshold value for the TR scheme can be interpreted in the context of hypothesis testing as the determination of a critical point (Beielstein & Markon 2001). The critical point $c_{1-\alpha}$ for a hypothesis test is a threshold to which one compares the value of the test statistic in a sample. It specifies the critical region CR and can be used to determine whether or not the null hypothesis is rejected. We are seeking a value $c_{1-\alpha}$, so that

$$\Pr\{S > c_{1-\alpha} \quad |H \text{ true }\} \leq \alpha, \tag{8.15}$$

where S denotes the test statistic, and the null hypothesis H reads: "There is no difference in means." Note that Eq. (8.15) is related to Eq. (3.6) in Sect. 3.1.2. The threshold acceptance selection method can be interpreted in a similar manner.

Generally, hypothesis testing interpreted as an automatic rule as introduced in Sect. 2.5.2 considers two-decision problems in which a null hypothesis H is either accepted or rejected. A false null hypothesis can be rejected 50% of the time by simply tossing a coin. Every time that heads comes up, H is rejected. The rejection procedures considered so far (Figs. 8.1–8.3) can be applied to s-decision problems. Here, larger sample sizes are required than for the two-decision problem. The probability of a correct selection for $s > 2$ is smaller than 50% if the decision is based on the roll of a fair s-sided die. To avoid too large sample sizes r for fixed s, the indifference zone δ^* can be increased, or the probability of a correct selection P^* can be reduced.

Known Theoretical Results

The theoretical analysis in Markon et al. (2001), where threshold rejection was introduced for evolutionary algorithms with noisy function values, was based on the progress rate theory on the sphere model and was shown for the $(1+1)$-ES. However, this theoretical result is only applicable when the distance to the optimum and the noise level are known—conditions that are not very often met in practice. By interpreting this result qualitatively, we can see that the threshold value τ should be increased while approaching the optimizer x^* (τ should be infinite when the optimum is obtained).

Another approach was used by Beielstein & Markon (2001). They demonstrated theoretically and experimentally how threshold rejection can improve the quality gain. This performance measure was introduced in Eq. (7.4) as the expected change in the function value. The influence of TR on the selection process was analyzed using a simple stochastic search model that is related to models proposed by Goldberg (1989) and Rudolph (1997b). This model possesses many crucial features of real-world optimization problems, i.e., a small probability of generating a better offspring in an uncertain environment. Then the search can be misled, although the algorithm selects only "better" candidates. TR can prevent this effect. In the simple stochastic search model the optimal threshold value could be calculated as a function of the noise strength, the probability of generating a better candidate, and the difference between the expectation of the function values of two adjacent states.

However, the determination of an optimal threshold value in this simple search model requires information that is usually not available and can only be estimated in real-world situations. For example, the probability of generating a better offspring is unknown during the search process; it is estimated in evolution strategies, cf. the definition of the estimated success rate in Eq. (8.19).

8.1.5 Sequential Selection

The selection procedures presented above can be extended to sequential strategies. We list some promising approaches that might be applicable as population based selection schemes.

For the case of unknown variance σ_ϵ^2, Santner (1976) proposed a two-stage selection scheme. Considering candidates with different means, unknown and not necessarily equal variances, Sullivan & Wilson (1984, 1989) presented a bounded subset selection for selecting a δ^*-near-best candidate.

A fully sequential procedure was proposed by Paulson (1964). Hartmann (1988, 1991) improved this procedure. Kim & Nelson (2001) extended the approach from Hartmann (1991) to unequal and unknown variances.

Bechhofer et al. (1990) and Kim & Nelson (2001) demonstrated the superiority of sequential selection methods over two-stage ranking-and-selection procedures. Pichitlamken & Nelson (2001) and Pichitlamken et al. (2003) presented a *sequential selection procedure with memory* (SSM). SSM is fully sequential with elimination: If inferior solutions are detected, they are eliminated immediately.

Bartz-Beielstein et al. (2005a) discussed the performance of the particle swarm optimization on functions disturbed by additive and multiplicative Gaussian noise. By comparing two simple algorithmic variants, they examine whether noise is more detrimental if it affects the selection of the local best or the selection of the global best. It is shown that parameter tuning alone is not sufficient to cope with noise, and that multiple sampling of solutions may be necessary to guarantee convergence. Finally, a new PSO variant with integrated sequential sampling technique is proposed, the *optimal computing budget allocation* (OCBA). OCBA was introduced in Chen et al. (1997). The aim of this method is to find the best within a set of candidate solutions and to select it with a high probability. The OCBA variant is a closed procedure and can be used for unknown and unequal variances. It draws samples sequentially until the computational budget is exhausted while adjusting the selection of samples to maximize the probability of a correct selection. The reader is referred to Chen et al. (2003) for a more detailed description. It is demonstrated that PSO with OCBA is superior to either the simple PSO or the PSO which samples each solution equally often. To show its superiority, the new PSO variant has to be tuned with SPO in advance.

8.2 Case Study I: How to Implement the (1 + 1)-ES

After introducing several strategies to cope with noise during selection, we demonstrate how the new experimentalism—and especially sequential parameter optimization—can be applied to analyze the effects of implementing these strategies. We have to define a baseline to judge the effect of a new selection operator. The (1+1)-ES is well suited to define this standard for comparisons.

It was developed in the 1960s as a minimal concept for an imitation of organic evolution (Schwefel 1995), therefore new algorithms should outperform this algorithm.

We will also consider an ES-variant that does not modify the step size $\sigma^{(t)}$ (cf. Eq. (6.3)). It is expected to be outperformed by other algorithms. However, sometimes unexpected results may occur. Probably nothing unexpected may happen, "but if something did happen, that would be a stupendous discovery" (Hacking 1983). This algorithm requires the specification of a (constant) step size σ value only.

Before we can analyze the performance of the (1+1)-ES, we have to choose an implementation of the control of the step size (mutation strength) σ. There are some open questions regarding this implementation. Wrongly implemented algorithms might lead to comparisons that are worthless. Step-size adaptation relies on the following heuristic: *The step size (standard deviation) should be adapted during the search. It should be increased if many successes occur, otherwise it should be reduced.* Rechenberg (1973) derived the 1/5 success rule, which was presented in Sect. 6.2.1, to control the step sizes. While discussing the 1/5 rule in evolution strategies, Schwefel (1995) notes:

> In many problems this [1/5 success rule] rule proves to be very effective in maintaining approximately the highest possible rate of progress towards the optimum. While in the rightangled corridor model the variances are adjusted once and for all in accordance with this rule and subsequently remain constant, in the sphere model they must steadily become smaller. The question then arises *(1) as how often* the success criterion should be tested and *(2) by what factor* the variances are most efficiently reduced or increased (p. 112; emphasis and numeration added).

To answer the first question ("how often"), Schwefel (1995) recommended using a measurement period $s_u = d$ and an adaptation interval size $s_n = 10d$.

To answer the second question ("by what factor"), Schwefel proceeds as follows: Let r denote the current (Euclidean) distance of the search point from the optimizer and d the problem dimension. The progress rate φ, which was introduced as performance measure (PM-7.14) in Sect. 7.2.3, is defined as the change in the distance to the optimizer. Based on the maximum progress rate,

$$\varphi_{\max} = k_1 r/d, \qquad k_1 \simeq 0.2025, \tag{8.16}$$

with a common variance σ^2, which is always optimal given by

$$\sigma_{\text{opt}} = k_2 r/d, \qquad k_2 \simeq 1.224, \tag{8.17}$$

for all components z_i of the random vector z (Eq.(6.1)), the following relation can be derived:

$$\lim_{d\to\infty} \frac{\sigma_{\text{opt}}^{(t+d)}}{\sigma_{\text{opt}}^{(t)}} = \lim_{d\to\infty} \left(1 - \frac{k_1}{d}\right)^d = \exp^{-k_1} \simeq 0.817 \simeq \frac{1}{1.224}. \tag{8.18}$$

An implementation of this heuristic with the design variables presented in Table 6.2 can be described as in Fig. 8.5.

Heuristic: The 1/5 success rule.
After every s_u $(= d)$ iterations, check how many successes have occurred over the preceding s_n $(= 10d)$ iterations. If this number is less than $s_n \times s_r$ $(= 2d)$, multiply the step lengths by the factor s_a $(= 0.85)$; divide them by s_a $(= 0.85)$ if more than $s_n \times s_r$ $(= 2d)$ successes occurred.

Fig. 8.5. Step-size adaptation. Settings from (Schwefel 1995) are shown in brackets; d denotes the problem dimension

We implemented the 1/5 rule as follows: A success vector $v^{(t)} \in \mathbb{B}^{s_n}$ is initialized at iteration $t = 1$: $v_k^{(t)} = 0,\ 1 \leq k \leq s_n$. If a successful mutation occurs at iteration t, the $(1+t \mod s_n)$th bit is set to 1, otherwise it is set to 0. After an initialization phase of s_n iterations, the success rate is estimated every s_u iterations as

$$\hat{s}_r^{(t)} = \sum_{k=1}^{s_n} v_k^{(t)} / s_n. \tag{8.19}$$

Implications arising from very small s_n values have to be mentioned, e.g., if $s_n = 1$, then the value of the success rate s_r has no effect on the algorithm. As $v^{(t)}$ stores information about previous successes and failures, it will be referred to as the *memory vector*.

8.2.1 The Problem Design Sphere I

The problem design $x^{(0)}_{\texttt{sphere}}$ (Table 8.1) was chosen for the first experiments from this case study. Schwefel (1988) notes that in lower dimension, nearly every strategy may achieve good results. To avoid improper conclusions, the problem dimension was taken as large as $d = 30$. To keep the (1 + 1)-ES competitive to other algorithms, we have chosen a budget of 1000 function

Table 8.1. Problem design $x^{(0)}_{\texttt{sphere}}$ for the experiments performed in this chapter. The experiment's name, the number of runs n, the maximum number of function evaluations $t_{\max}$, the problem's dimension d, the initialization method, the termination criterion, the starting point for the initialization of the object variables $x^{(0)}$, as well as the optimization problem and the performance measure (PM) are reported

Design	Init.	Term.	PM	n	$t_{\max}$	d	$x^{(0)}$
$x^{(1)}_{\texttt{sphere}}$	DETEQ	EXH	MBST	50	1000	30	100

evaluations only. Note that different settings, especially for large $t_{\max}$ values, might produce directly opposed results. Anyhow, in many real-world applications more than 1000 function evaluations are prohibitive. The deterministic initialization scheme DETEQ, which was presented in Sect. 5.5.1, has been selected to reduce the variance and to enable the application of the fundamental ANOVA principle (Sect. 3.4).

The mean best function value (Sect. 7.2.2) has been chosen as a performance measure because it is commonly used. Note that we have chosen the problem design $x^{(0)}_{\texttt{sphere}}$ (Table 8.1) as a starting point for further investigations. Settings from this design will be varied during the experimental analysis. Algorithm design $x^{(0)}_{\text{ES}}$ (Table 8.2) has been used for the first experiments.

Design plots give an overview how the algorithm behaves. They require factorial designs and can be used to determine ranges for suitable parameter settings. However, design plots do not provide any information about factor interactions. The design plots (Fig. 8.6) suggest that small values, e.g., $s_n = 10$, should be chosen for the adaptation interval, a success rate of $1/5$, a step-size adaptation value of 0.85, and values in the range from 30 to 50 for the update interval s_u result in an improved performance of the $(1+1)$-ES on the 30-dimensional sphere function.

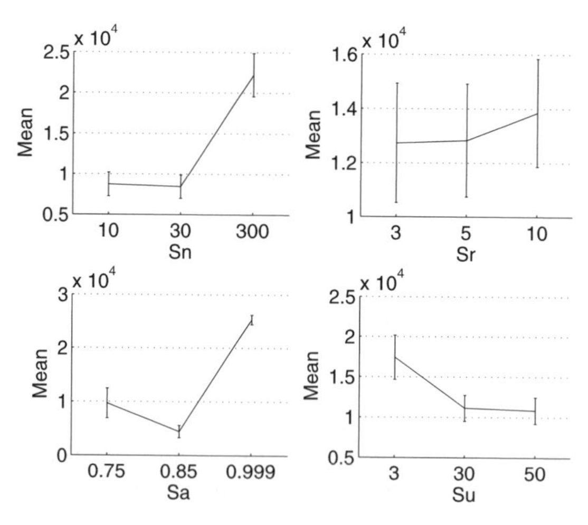

Fig. 8.6. Design plots from the first experiments that use problem design $x^{(0)}_{\texttt{sphere}}$ and algorithm design $x^{(0)}_{\text{ES}}$. They suggest that the value of 300 for the adaptation interval is much too high; smaller values for the step size adaptation, and larger values for the update interval, respectively, appear to be beneficial

This first picture can be misleading if interactions are omitted. Interaction plots—which use the same data as the design plots—are considered next. Combining the results from both plots (Figs. 8.6 and 8.7), we can conclude

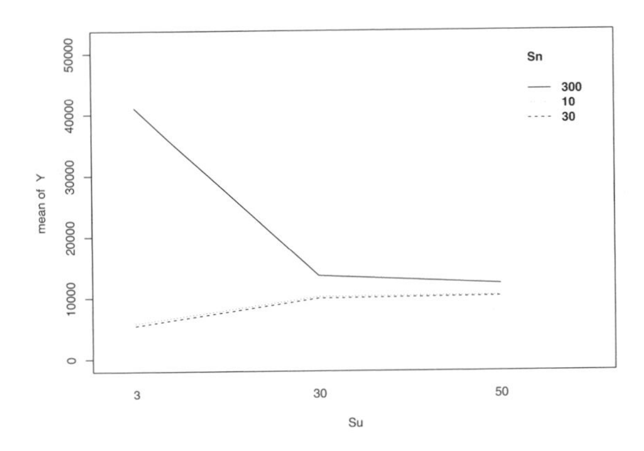

Fig. 8.7. Interaction plots provide important information to complement interpretations of design plots (Fig. 8.6). Two conclusions can be drawn from this figure: (1) Lower values for the adaptation interval improve the algorithm's performance. (2) This figure illustrates nicely how interactions can misguide the analysis. Interactions caused by a worse setting of the adaptation interval ($s_n = 300$) cause the impression (Fig. 8.6) that small s_n values should be avoided. However, as the interaction plot reveals, just the opposite is true: smaller values are better

that the role of the adaptation interval has to be reconsidered: small values, e.g., $s_u = 3$, improve the algorithm's performance.

After further design points have been added to refine the analysis (algorithm design $x_{\mathrm{ES}}^{(1)}$ in Table 8.2), an unexpected effect occurs. The algorithm performs significantly better if even values for s_n are chosen.

This effect could be observed if a success rate of 1/2 was chosen. It can be explained as follows. There are four possible settings for the memory vector if $s_n = 2$: $v \in \{(0,0),(0,1),(1,0),(1,1)\}$. The memory vector is used to store the number of successes. The step-size adaptation scheme was implemented as shown on the left in Fig. 8.8: Decrease the step size, if

Table 8.2. Algorithm designs for the (1 + 1)-ES. The adaptation interval s_n, the success rate $1/s_r$, the step-size adjustment factor s_a, and the update interval s_u were defined in Table 6.2. An initial step-size value $\sigma^{(0)} = 1$ was chosen for these experiments. Note that we use a compact representation for algorithm designs in this and the following tables, i.e., 2:9 is the set of integers from 2 to 9.

Design	s_n	s_r	s_a	s_u
$x_{\mathrm{ES}}^{(0)}$	{10,30,300}	{3,5,10}	{0.75,0.85,0.999}	{3,30,50}
$x_{\mathrm{ES}}^{(1)}$	2:9	2:5	0.75:1:0.95	{2,10,20}

```
if iter > sn
 if (mod(iter ,su)==0)
  if sum(v)/sn < 1/sr
   s = s*sa;
  else
   s = s/sa;
  end
 end
end
```

```
1 if iter > sn
   if (mod(iter ,su)==0)
3   if sum(v)/sn < 1/sr
     s = s*sa;
5   elseif sum(v)/sn > 1/sr
     s = s/sa;
7   end
   end
9 end
```

Fig. 8.8. Implementation of the (1 + 1)-ES step size adaptation; v denotes the memory vector, s_n, s_r, s_a, and s_u as defined in Table 6.2. *Line 5* has been modified to avoid bias

$$\sum_{i=1}^{s_n} v_i s_n < 1/s_r. \tag{8.20}$$

Now consider $s_r = 2$. If the distance to the optimizer is sufficiently large, Eq. (8.20) is true in one out four cases only—the step-size adaptation is biased. Increasing step sizes are preferred. This bias enables the algorithm to accelerate the search in the beginning, as can be seen in Fig. 8.9 on the left. A similar argument shows that the bias is negligible for even s_n values, which leads to a decrease in performance (right panel in Fig. 8.9). Hence, the algorithm is not able to find a vicinity of the optimum with the prespecified budget of function evaluations due to too small step sizes (Fig. 8.10).

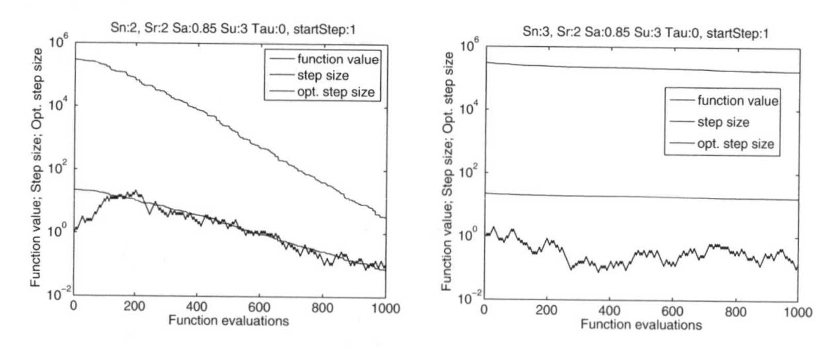

Fig. 8.9. Function values and step sizes. *Left*: The size of the memory vector is $s_n = 2$. The step sizes are increased during the first phase of the search. *Right*: The size of the memory vector is $s_n = 3$. From *top to bottom*: function values, optimal step size, and step size. No sufficient increase in the step sizes occurs

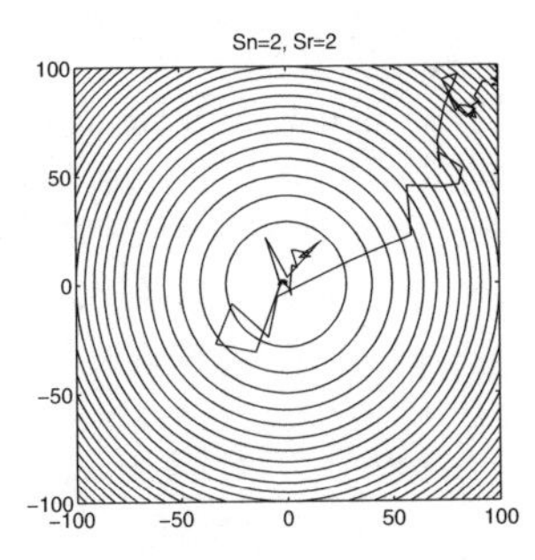

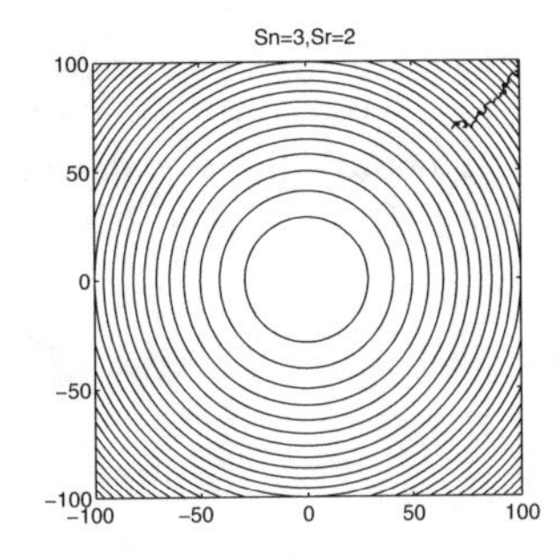

Fig. 8.10. Visualization. *Left*: The step size is increased in the first phase. Approaching the optimizer, it is reduced. *Right*: The step size cannot be increased; it remains too small and prevents convergence

This effect occurs only for a success rate value $s_r = 2$. After the algorithm was modified as shown in Fig. 8.8, the bias could be avoided and the effect disappeared. Hence, implementation details can have a significant influence on the algorithm's performance. For example, replacing "$<$" with "$\leq$" can have a significant impact on the performance of search heuristics: Jansen & Wegener (2000) compared the $(1+1)$-EA to a variant, the so-called $(1+1)^*$-EA, which accepts only the offspring whose function value is strictly better (smaller) than the function value of its parent.

Furthermore, the first experiments show that under the assumptions specified in the problem design $x^{(0)}_{\texttt{sphere}}$ (e.g., moderate initial step-size values and starting points are not in the direct vicinity of the optimizer), the success rate should be smaller than 0.5 to enable a proper increase in the step-size values during the first phase of the search. We continue our search for good parameter settings of the $(1 + 1)$-ES. Algorithm design $x^{(4)}_{\mathrm{ES}}$ (Table 8.4) was used for these experiments.

A memory vector of size 3 leads to worse results. This effect occurs independently from the starting value $\sigma^{(0)}$. It can be explained by considering the changes in the step sizes for different s_r values for a memory vector of length $s_n = 3$ (Table 8.3).

The probability of decreasing step sizes is too high if $s_r = 2$, an s_r value of 3 performs best, and $s_r = 4$ results in step sizes that are too large, respectively. A setting that enables a decrease in the step sizes, but prefers an increase if successes occur, works best.

Similar considerations lead to the following result for $s_n = 4$. If $s_r = 3$, then 6 (from $2^4 = 16$) configurations decrease the step size. If $s_r = 4$ only one configuration decreases the step size, four are neutral, and 11 increase the step size, and if $s_r = 5$, 15 lead to an increase, and only one configuration leads to a decrease. Therefore, $s_r = 4$ seems to be a good choice if a memory vector with $s_n = 4$ is chosen.

Table 8.3. Effects of different s_r values for a memory vector of length 3. The symbol "−" denotes that the step size will be decreased, "o" denotes no change, and "+" an increase. The probability of decreasing step sizes is too high if $s_r = 2$, a s_r value of 3 performs best, and $s_r = 2$ leads to too small, $s_r = 4$ to too large step sizes, respectively

v_1	v_2	v_3	$s_r = 2$	$s_r = 3$	$s_r = 4$
0	0	0	−	−	−
0	0	1	−	o	+
0	1	0	−	o	+
0	1	1	+	+	+
1	0	0	−	o	+
1	0	1	+	+	+
1	1	0	+	+	+
1	1	1	+	+	+

This knowledge can be used to predict the behavior of the (1 + 1)-ES. Consider, for example, $s_n = 3$: We expect a decrease in the step sizes if $s_r = 2$; the algorithm gets stuck before reaching the optimum. A value of $s_r = 4$ might result in step sizes that are too large, and we expect that a value of $s_r = 3$ works fine. These assumption have been confirmed experimentally, as can be seen in Fig. 8.11.

These experiments conclude the preexperimental planning phase (cf. step (S-1), Sect. 7.5). Now we can formulate a scientific claim.

(C-8.1) It is better to choose a (1+1)-ES with a small memory vector ($s_n \ll 10d$) if the number of function evaluations is restricted.

SPO as introduced in Sect. 7.5 can be applied next. It takes the dependency of the algorithm's performance on suitable chosen parameter values into account. It can discover implementation details that are not predictable by theory. Therefore we recommend SPO to detect good parameter settings at this stage of the experimentation. Note that classical DOE techniques enable the user to specify certain design points, whereas DACE sets design points randomly in the latter. The user has to specify regions of interest.

Algorithm design $x_{\text{ES}}^{(5)}$ from Table 8.4 was chosen to generate a Latin hypercube design with 80 design points. Each design point was evaluated 10 times, resulting in 800 algorithm runs during the first phase. As a result of the SPO procedure, the improved design point x_{ES}^* (Table 8.4) has been detected.

The scientific thesis is broken down into several statistical hypotheses, e.g.,

(H-8.1) The (1 + 1)-ES that uses algorithm design x_{ES}^* (Table 8.4) performs significantly better than the (1 + 1)-ES with default parameterization $x_{\text{ES}}^{(\text{def})}$, if problem design $x_{\texttt{sphere}}^{(0)}$ is considered.

The final comparison is based on $n = 500$ runs. OSL plots and a comparison of the numerical data in Table 8.5 show that there is a difference in

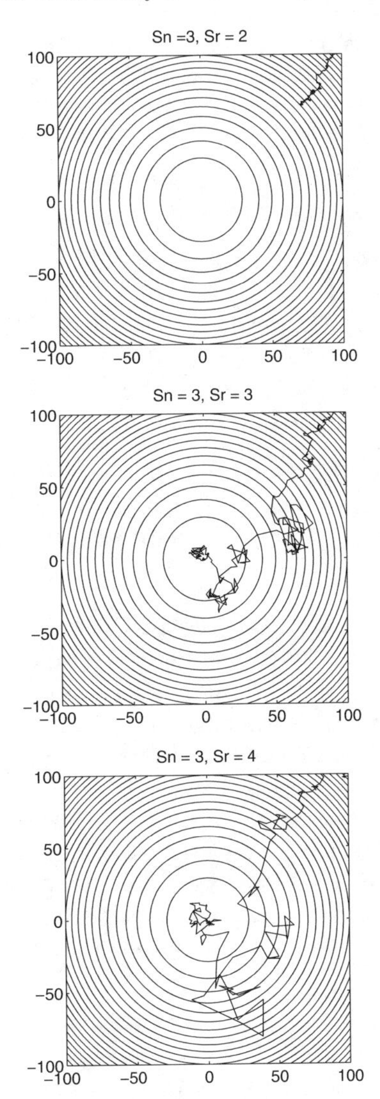

Fig. 8.11. Visualization of the position of the candidate solution in the search space. Only two dimensions are considered. A memory vector with $s_n = 3$ was used. Problem design $x^{(0)}_{\texttt{sphere}}$ and algorithm design $x^{(4)}_{\mathrm{ES}}$. The graphs show (from *top to bottom*) runs with success rates 1/2, 1/3, and 1/4, respectively. The success rate 1/3 leads to the best results. These graphs support the hypotheses from above and the considerations from Table 8.3

Table 8.4. Algorithm designs for the (1 + 1)-ES. The adaptation interval s_n, the success rate $1/s_r$, the step-size adjustment factor s_a, and the update interval s_u were defined in Table 6.2. An initial step-size value $\sigma^{(0)} = 1$ was chosen for these experiments. The values for the default (1 + 1)-ES were motivated in Fig. 8.5

Design	s_n	s_r	s_a	s_u
$x_{\text{ES}}^{(4)}$	3	2:4	0.85	10
$x_{\text{ES}}^{(\text{def})}$	300	5	0.85	30
$x_{\text{ES}}^{(5)}$	5:30	3:8	0.75:1:0.9	2:20
x_{ES}^{*}	9	3.97	0.78	11

Table 8.5. Experimental results from (1 + 1)-ES with default and improved algorithm designs. Problem design $x_{\text{sphere}}^{(0)}$ was used

Algorithm	Mean	Median	SD	Min	Max	$\min_{\text{boot}}$
$x_{\text{ES}}^{(\text{def})}$	4436	4324	1322	1422	10253	3413
x_{ES}^{*}	**4.04**	**2.90**	**3.97**	**0.14**	**32.22**	**1.60**

means between these two algorithm designs. The difference is so overwhelming, no further statistical tests are necessary to validate the hypothesis that x_{ES}^{*} performs better than $x_{\text{ES}}^{(\text{def})}$ on the 30 dimensional sphere, if only a budget of 1000 function evaluations is available.

To illustrate why a smaller memory vector improves the performance we plotted function values, step sizes, and optimal step sizes (cf. Eq. (8.17)). The results from Fig. 8.12 demonstrate that (1+1)-ES is a good hill-climber, that means it makes small steps, if small steps are advantageous, and large steps, if large steps are better. Therefore, a large memory vector is obstructive for this problem instance, because it prevents a flexible adaptation of the step sizes. Furthermore, the experiments confirmed the 1/5 rule.

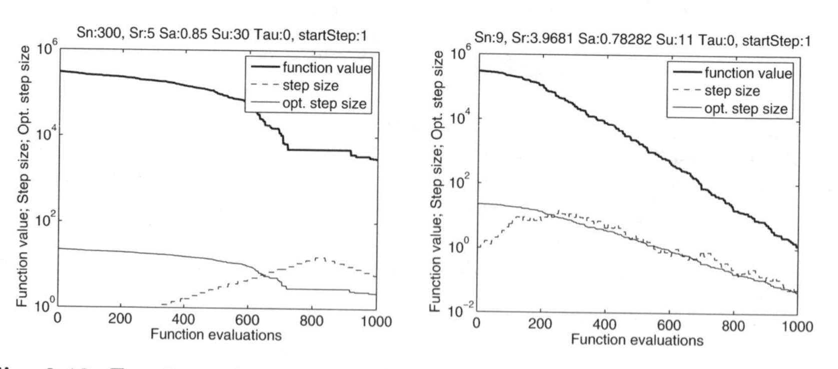

Fig. 8.12. Function values, step size, and optimal step size. *Left*: default setting $x_{\text{ES}}^{(\text{def})}$, *right*: improved setting x_{ES}^{*}

Summarizing, we can state that a working (1 + 1)-ES that performs significantly better than the standard (1 + 1)-ES was found. We claim that this result is scientifically meaningful, and, as a consequence, we recommend not to use the default algorithm design $x_{\text{ES}}^{(\text{def})}$ with $s_n = 10d$, if the number of available function evaluations is low. As an important consequence, which has an impact that goes beyond this case study, we can state that results from algorithm comparisons become questionable, if the in-between variance, which is caused by different parameter settings for one algorithm, is already higher than the between-group variance.

We have demonstrated how a baseline for comparisons—especially for stochastic search heuristics—can be defined. SPO or related techniques should be applied in the first phase of these comparisons.

8.3 Case Study II: The Effect of Thresholding

The first case study was devoted to the question how to define a standard that can be used for comparisons. Now we will extend this default algorithm by implementing additional selection schemes. This procedure is referred to in other contexts as *hybridization*.

Minimization of the d-dimensional sphere function $\sum_{i=1}^{d} x_i^2$ (Table 4.1), disturbed by additive Gaussian noise $\epsilon \sim \mathcal{N}(0, \sigma_\epsilon^2)$, is considered. To analyze the influence of three selection schemes on the performance of the (1 + 1)-ES, three algorithms are compared:

1. (1 + 1)-ES
2. TR, that is, the (1 + 1)-ES with positive τ values
3. TA, that is, the (1 + 1)-ES with negative τ values

Figure 8.13 illustrates the hybrid ES/TS algorithm.

Bartz-Beielstein (2005a) demonstrated that constant threshold values worsen the performance of the (1 + 1) on the sphere if the function value can be determined exactly. Here we will consider the noisy sphere.

8.3.1 Local Performance

Instead of analyzing the global performance, e.g., the mean function values after 1000 function evaluations, we start with an analysis of the local performance. Simulations were performed to analyze the influence of the threshold value on the progress rate φ and on the success rate $1/s_r$ as defined in Sect. 7.2. Figure 8.14 describes the simulation procedure. The problem designs are shown in Table 8.6. The corresponding hypothesis reads:

(H-8.2) Threshold selection produces better quality results than plus selection if the function values are disturbed by additive, Gaussian noise. The results are independent of the test problem.

Procedure: (1 + 1)-ES/TS

Initialization: Initialize the iteration counter: $t = 1$. Determine: (i) a point $X_1^{(t)}$ with associated position vector $x_1^{(t)} \in \mathbb{R}^d$, (ii) a standard deviation $\sigma^{(t)}$, and (iii) a threshold value $\tau^{(t)}$. Determine the function value $y_1 = f(x_1^{(t)})$.

`while` some stopping criterion is not fulfilled `do`

`repeat` M times:

Mutation: Generate a new point $X_2^{(t)}$ with associated position vector $x_2^{(t)}$ as follows:

$$x_2^{(t)} = x_1^{(t)} + z, \tag{8.21}$$

where z is a d-dimensional vector. Each component of z is the realization of a normal random variable Z with mean zero and standard deviation $\sigma^{(t)}$.

Evaluation: Determine the function value $y_2 = f(x_2^{(t)})$.

Selection: Accept $X_2^{(t)}$ as $X_1^{(t+1)}$ if

$$y_2 + \tau^{(t)} < y_1, \tag{8.22}$$

otherwise retain $X_1^{(t)}$ as $X_1^{(t+1)}$. Increment t.

`end.`

Adaptation:

$$\text{Update} \quad \sigma^{(t)}. \quad \text{Update} \quad \tau^{(t)}. \tag{8.23}$$

`done.`

Fig. 8.13. The hybrid evolution/threshold selection strategy (ES/TS). The two-membered evolution strategy or (1+1)-ES for real-valued search spaces uses $M = 1$ and $\tau^{(t)} \equiv 0$. The symbol f denotes an objective function $f : \mathbb{R}^d \to \mathbb{R}$ to be minimized. Threshold selection uses a constant step size $\sigma^{(t)} \equiv \sigma$ and a threshold adaptation scheme

Table 8.6. Problem designs for the (1 + 1)-ES simulation runs. The progress rate PRATE was chosen as a performance measure. Note that the sample size is denoted by r. The starting point is chosen deterministically: $x^{(0)} = 1$

Design	r	$t_{\max}$	d	Init.	Term.	$x^{(0)}$	Perf.	Noise
$x^{(1)}_{\texttt{abs}}$	10^5	1	1	DETEQ	EXH	1	PRATE	1
$x^{(1)}_{\texttt{id}}$	10^5	1	1	DETEQ	EXH	1	PRATE	1
$x^{(1)}_{\texttt{sphere}}$	10^5	1	1	DETEQ	EXH	1	PRATE	1

Procedure: (1+1)-ES simulation to approximate the one-generation progress φ

Initialization: Initialize the sample counter $i = 1$. The index i has been suppressed to improve readability. Choose one initial parent X_1 with associated position vector $x_1 \in \mathbb{R}^d$. Choose the standard deviation $\sigma \in \mathbb{R}_+$, the threshold value $\tau \in \mathbb{R}$, and the noise level σ_ϵ^2.

`repeat`

Mutation: Generate a new point X_2 with position vector x_2 as follows:

$$x_2 = x_1 + z, \tag{8.24}$$

where each component of the d-dimensional vector z is the realization of random variable $Z \sim \mathcal{N}(0, \sigma^2)$.

Evaluation: Determine the function values

$$y_j = f(x_j), \tag{8.25}$$

$$\tilde{y}_j = f(x_j) + w_j, \tag{8.26}$$

where w_j are realizations of $\mathcal{N}(0, \sigma_\epsilon^2)$ distributed random variables W_j, $j = 1, 2$.

Selection: Accept X_2 if

$$\tilde{y}_2 + \tau < \tilde{y}_1, \tag{8.27}$$

otherwise reject X_2.

Progress: Determine

$$\delta_i = \begin{cases} x_1 - x_2, & \text{if } X_2 \text{ was accepted,} \\ 0, & \text{otherwise.} \end{cases} \tag{8.28}$$

Increment i.

`until` r samples have been obtained.

Return $\sum_{i=1}^{r} \delta_i / r$, an estimate of the expected progress φ from generation g to $g + 1$, see (PM-7.14) in Sect. 7.2.

Fig. 8.14. $(1+1)$-ES simulation to study the effect of threshold selection on the progress rate φ

To reject hypothesis (H-8.2), we have to find a test function on which the (1+1)-ES performs better than the (1+1)-TS. We consider three candidates: the absolute value function (`abs`), the identity function (`id`), and the sphere function (`sphere`). A constant noise level $\sigma_\epsilon^2 = 1$, the starting point $x^{(0)} = 1$, and the step size $\sigma = 1$ have been chosen for these experiments. Figure 8.15 illustrates the results.

The approximated progress rate φ is plotted against the threshold value τ. Positive φ values are better, because φ is the expected change in the distance of

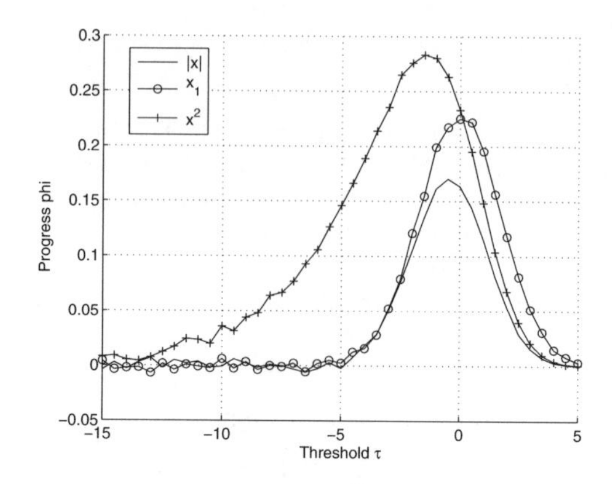

Fig. 8.15. Three different functions: `abs`, `id`, and `sphere` and related problem designs $x^{(1)}_{\texttt{abs}}$, $x^{(1)}_{\texttt{id}}$, and $x^{(1)}_{\texttt{sphere}}$, respectively, from Table 8.6. Results from the simulation study described in Fig. 8.14 to analyze the effect of TS on the progress rate. Progress rate φ plotted against threshold value τ. Noise $\sigma^2_\epsilon = 1$, starting point $x^{(0)} = 1$, and step size $\sigma = 1$. These results indicate that threshold selection produces worse results on the identity function, whereas positive effects could be observed on $x^{(1)}_{\texttt{abs}}$ and $x^{(1)}_{\texttt{sphere}}$

the search point to the optimum in one generation. The results from this study show that threshold acceptance ($\tau \leq 0$) can improve the progress rate on the absolute value function and on the sphere function. But threshold acceptance worsens the performance on the identity function (`id`). And threshold rejection ($\tau > 0$) worsens the progress rate in any case.

What are the differences between `id`, `abs`, and `sphere`? The starting point $x^{(0)}$ was chosen in the immediate vicinity of the global minimizer x^* of the test functions `abs` and `sphere`. This closeness to the optimum might explain the effect of the threshold selection scheme. This consideration leads to the next hypothesis:

(H-8.3) Threshold selection produces better quality results than plus selection in the vicinity of the global minimizer x^*, if the function values are disturbed by additive, Gaussian noise. The results are independent of the test problem.

The problem design in Table 8.7 was used to perform the experiments. As before, the simulation procedure shown in Fig. 8.14 was used to approximate the one-step progress rate φ.

The results (not shown here) indicate that the starting point $x^{(0)}$ influences the threshold selection scheme. The optimal threshold value decreases (becomes negative) as the distance of the starting point $x^{(0)}$ to the optimum x^* is reduced. A similar effect could be observed for the absolute value func-

Table 8.7. Problem design for the (1 + 1)-ES simulation runs. The distance to the optimum of the starting point is varied

Design	r	$t_{\max}$	d	Init.	Term.	x_l	x_u	Perf.	Noise
$x^{(2)}_{\texttt{abs}}$	10^5	1	1	DETMOD	EXH	1	10	PRATE	1
$x^{(2)}_{\texttt{id}}$	10^5	1	1	DETMOD	EXH	1	10	PRATE	1
$x^{(2)}_{\texttt{sphere}}$	10^5	1	1	DETMOD	EXH	1	10	PRATE	1

tion `abs`. The influence of the TS scheme vanishes if the starting point $x^{(0)} = 2$ was chosen.

Both functions, `abs` and `sphere`, are convex. Recall that a function $f(x)$ is convex on an interval $[a, b]$ if for any two points x_1 and x_2 in $[a, b]$,

$$f\left(\frac{1}{2}(x_1 + x_2)\right) \leq \frac{1}{2}f(x_1 + x_2).$$

A function $f(x)$ is strictly convex if $f\left(\frac{1}{2}(x_1 + x_2)\right) < \frac{1}{2}f(x_1 + x_2)$. That is, a function is convex if and only if its epigraph (the set of points lying on or above the graph) is a convex set. The sphere function x^2 and the absolute value function $|x|$ are convex. The function $\texttt{id}(x) = x$ is convex but not strictly convex. Thus, the next hypothesis reads:

(H-8.4) Let f denote a strictly convex test function. Threshold acceptance produces better quality results than plus selection in the vicinity of the global minimizer x^* of f if the function values are disturbed by additive, Gaussian noise.

To test hypothesis (H-8.4), we simulate the (1 + 1)-ES on the bisecting line cosine function (`bilcos`). This function has infinitely many local minimizers $x_i = 2i + \epsilon$ and infinitely many local maximizers $x_i = 2i - 1 - \epsilon$, with $i \in \mathbb{Z}$ and $\epsilon = \sin^{-1}(1/\pi)/\pi \approx -.1031$.

Figure 8.16 illustrates the results from these simulations: The (1 + 1)-ES with threshold acceptance performs better with threshold acceptance if a starting point $x^{(0)}$ in the neighborhood of a local minimum is chosen. Threshold rejection improves the approximated progress rate in the neighborhood of a local maximum. A zero threshold value is best if $x^{(0)}$ is placed between two local optima. This simulation study demonstrated that the curvature influences the optimal threshold value: The (1 + 1)-ES with threshold acceptance performs better on strictly convex functions than the (1 + 1)-ES, whereas the (1 + 1)-ES with threshold rejection performs better than the (1 + 1)-ES on strictly concave functions.

The next experiments to refine our hypothesis are conducted to analyze the influence of the noise level σ^2_ϵ. Therefore, we state:

(H-8.5) Let f (g) denote a strictly convex (concave) test function. Threshold acceptance (rejection) produces better quality results than plus selection

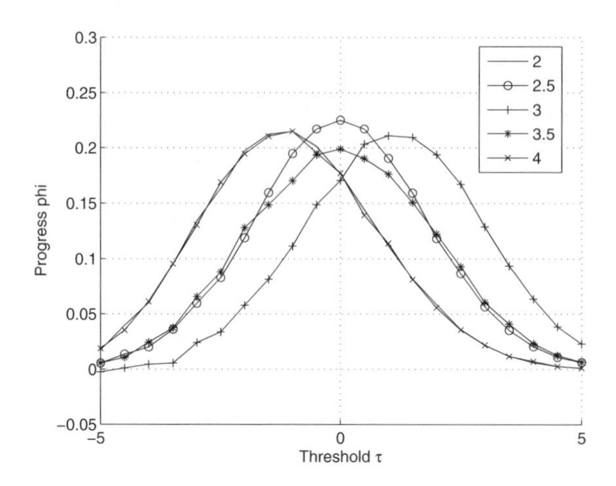

Fig. 8.16. Progress rate φ and threshold selection. Bisecting line cosine function `bilcos`. *Curves* represent results from experiments with different starting points $x^{(0)} \in \{2, 2.5, 3, 3.5, 4\}$. A positive threshold value improves the performance if the simulation is started from a local maximum, i.e., $x^{(0)} = 3$

in the vicinity of a local minimum (maximum) of f (g) if the function values are disturbed by additive, Gaussian noise. The optimal threshold value τ^* increases as the noise level σ_ϵ^2 grows.

Experiments varying the noise level σ_ϵ^2 (not presented here) gave no indication that (H-8.5) is wrong. If the noise level is very high, threshold selection cannot improve the progress rate. The influence of the TS is only marginal for small noise levels.

Global Performance

The analysis from the previous section considered the local performance of the $(1+1)$ algorithms. The progress rate PRATE was used to measure the one-generation improvement. To analyze the global performance, other performance measure will be used. A commonly used measure is the mean value from n runs. In addition to this measure, the median, standard deviation, minimum, maximum values, and the $\min_{\text{boot}}$ value will be reported.

Based on considerations developed by Markon et al. (2001), we implemented the threshold adaptation as

$$\tau^{(t)} = \frac{\sigma_\epsilon^2 d}{2 y_1^{(t)}}, \tag{8.29}$$

where d denotes the problem dimension, σ_ϵ the noise level, and $y_1^{(t)}$ the function value at time step t. The problem design $x_{\texttt{sphere}}^{(1)}$ (Table 8.8) was used in this study.

Table 8.8. Problem design $x^{(1)}_{\texttt{sphere}}$ for the experiments performed in this section. The experiment's name, the number of runs n, the maximum number of function evaluations $t_{\max}$, the problem's dimension d, the initialization method, the termination criterion, the starting point for the initialization of the object variables $x^{(0)}$, as well as the optimization problem, the performance measure (PM), and the noise level are reported

Design	Init.	Term.	PM	n	$t_{\max}$	d	$x^{(0)}$	Noise
$x^{(1)}_{\texttt{sphere}}$	DETEQ	EXH	MBST	50	1000	30	100	1

The scientific claim reads:

(C-8.2) Threshold rejection improves the performance of the $(1+1)$-ES on the noisy sphere.

We have to break this thesis down to formulate hypotheses that can be tested statistically. A related statistical hypothesis reads "Given problem design $x^{(1)}_{\texttt{sphere}}$ (Table 8.8), threshold rejection outperforms the $(1+1)$-ES and the threshold acceptance strategy."

We applied SPO to determine improved algorithm designs for each algorithm (Table 8.9): $x^{\star}_{\text{ES}}$ for the $(1+1)$-ES, $x^{\star}_{\text{TR}}$ for the threshold rejection strategy, and $x^{\star}_{\text{TA}}$ for the threshold acceptance strategy, respectively. A comparison based on these designs with $n = 500$ runs is performed. An inspection of the results (Table 8.10) shows that just the opposite of the expected behavior occurred: Threshold rejection worsens the performance of the algorithm, whereas threshold acceptance improves the performance with the exceptions of two performance measure.

An interpretation of the scientific meaning of these results is difficult. The analysis reveals that TA performs better than the $(1+1)$-ES on average, but the standard $(1+1)$-ES was able to detect lower function values (Table 8.10). Using the $\min_{\text{boot}}$ procedure, the difference becomes smaller. Histograms would also display these relation graphically.

Table 8.9. Improved algorithm designs for the $(1+1)$-ES, TR, and TA optimizing the noisy sphere: $x^{\star}_{\text{ES}}$, $x^{\star}_{\text{TR}}$, and $x^{\star}_{\text{TA}}$, respectively. The adaptation interval s_n, the success rate $1/s_r$, the step-size adjustment factor s_a, and the update interval s_u were defined in Table 6.2. An initial step-size value $\sigma^{(0)} = 1$ was chosen for these experiments

Design	s_n	s_r	s_a	s_u
$x^{\star}_{\text{ES}}$	15	4.06	0.79	10
$x^{\star}_{\text{TR}}$	10	4.76	0.85	11
$x^{\star}_{\text{TA}}$	12	5.52	0.72	12

Table 8.10. Experimental results from (1 + 1)-ES with default and improved algorithm designs. Problem design $x^{(1)}_{\text{sphere}}$ was used. These experimental data suggest that threshold acceptance has a positive influence on the performance of the (1+1)-ES, whereas threshold rejection worsens its performance. Best results are printed in *boldface*

Algorithm	Mean	Median	Sd	Min	Max	$\min_{\text{boot}}$
$x^{\star}_{\text{ES}}$	19.4	17.91	8.28	**5.33**	**69.28**	13.40
$x^{\star}_{\text{TR}}$	20.97	20.05	7.64	7.92	91.61	15.50
$x^{\star}_{\text{TA}}$	**17.56**	**15.69**	**7.16**	7.50	69.29	**13.09**

Figure 8.17 supports the hypothesis that TA can improve the performance of a (1 + 1)-ES in the vicinity of an optimum. We suggest that this result is caused by the following algorithm behavior:

1. As threshold acceptance increases the success rate, small step sizes are avoided. TA increases the success rate erroneously in some cases, and new candidates are accepted that are only seemingly better than the actual best. Recall that threshold acceptance increases the type-I error (Eq. 3.2).
2. Threshold rejection decreases the success rate, which leads to smaller step sizes. Even better candidates are sometimes rejected. Recall that threshold rejection reduces the error of the first kind.

Threshold acceptance as implemented here (cf. Eq. 8.29) has nearly no influence on the performance of the algorithm during the first phase of the search. Furthermore, OSL plots (Fig. 8.18) can provide valuable information for this interpretation. Finally, the experimenter has to interpret the optimization scenario—which specifies whether one good solution or solutions that are good on average are requested—to decide if threshold acceptance should be used.

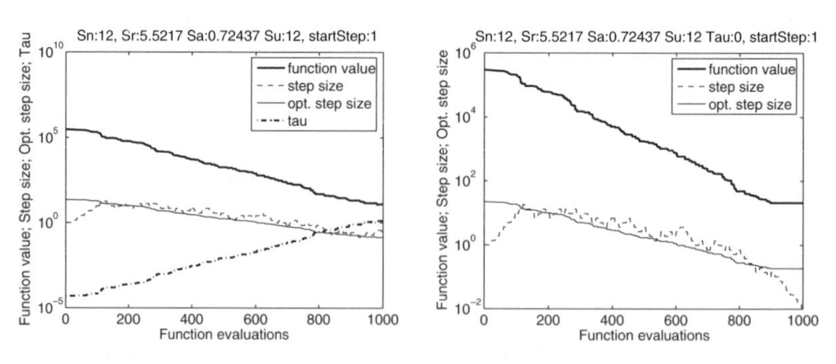

Fig. 8.17. Function values and step sizes. Threshold acceptance prevents small step sizes in the final phase of the (1 + 1)-ES run

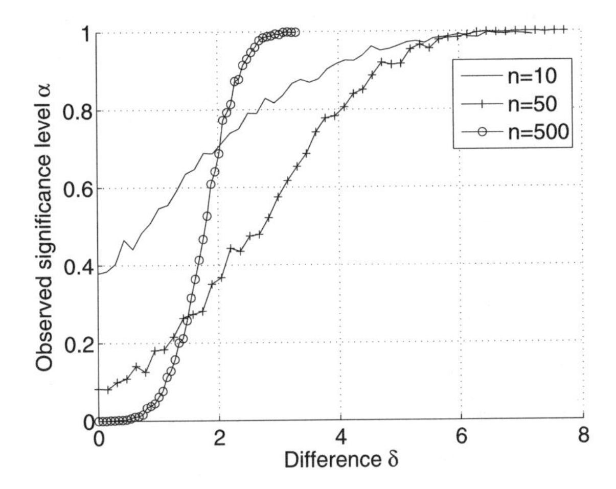

Fig. 8.18. OSL plot. Observed significance plot for the (1+1)-ES and the (1+1)-ES with threshold acceptance. There is a difference in means, but the observer has to decide whether this difference is significant or not. This is neither case RE-2.1 nor case RE-2.2, cf. Sect. 2.5.4

8.4 Bounded Rationality

Simon's (1955) concept of *bounded rationality* considers

1. cognitive limits of actual humans
2. environments that permit simplifications of rational decision making

Gigerenzer & Selten (2002) note that optimization is often based on uncertain assumptions (guesswork), and there maybe about as many different outcomes of optimization strategies as there are sets of assumptions: "In these real-world cases, it is possible that simple and robust heuristics can match or even outperform a specific optimization strategy." Imitation, equal weighting, take the best, take the first, and small-sample inferences are examples of fast and frugal heuristics (Goldstein et al. 2002). Another example, where a simple model outperforms a complex model, is given by Forster & Sober (1994).

Example 8.1 (Overfitting). Curve fitting in classical and modern regression analysis consists of two steps. A family of curves is selected first, i.e., linear, quadratic, or more sophisticated functions. Simple curves are preferred in this step; consider the situation depicted in Figure 8.19. In a second step the curve in that family that fits the data best is selected. To perform the second step some measure of goodness-of-fit is necessary.

Simplicity and goodness-of-fit are two conflicting goals. Therefore, the following question arises: Why should the simplicity of a curve have any relevance to our opinions about which curve is true? Including the prediction error to these considerations provides a deeper understanding. A result in statistics from Akaike shows how simplicity and goodness-of-fit contribute to a curve's expected accuracy in making predictions (Kieseppä 1997). The predictive power of the curve is more important than its fit of the actual data. Curves that fit a given data set perfectly will usually perform poorly when they are used to make predictions about new data sets, a phenomenon known as overfitting. ■

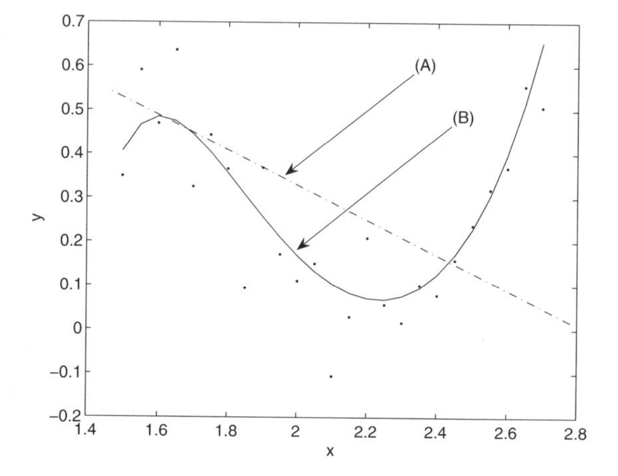

Fig. 8.19. Simplicity of curves. Linear (*A*) and cubic (*B*) curves. Imagine a third curve (C) that fits every data point. Why do scientists prefer curve (B)?

Threshold selection, the (1+1)-ES, and the 1/5 success rule can be classified as simple and robust heuristics. They avoid overfitting, because they only use a minimal amount of information from the environment. Under this perspective, algorithm tuning as introduced in Chap. 7 can be seen as an analogue to curve fitting (Chap. 3): Algorithms with more exogenous strategy parameters enable a greater flexibility for the cost of an expensive tuning procedure. Poorly designed algorithms can cause "overfitting"—they are able to solve one specific problem only. Domain-specific heuristics require a moderate amount of information from the environment.

Does the environment permit a reduction of the required amount of information for decision making or optimization? Although simple heuristics often work well, they can be misled easily if they are used in unsuitable environments. Consider, for example, the 1/5 rule: Its validity is not restricted to the `sphere` function. However, Beyer (2001) describes fitness landscapes in which the 1/5 rule fails, for example, when the objective function is not continuously differentiable in the neighborhood of the parental position vector. Recognizing the situations in which domain-specific heuristics perform better than other strategies provides a better understanding of their mechanisms. Understanding is seen here as to figure out in which environments a simple tool can match or even outperform more complex tools. Gigerenzer et al. (1999) use the term "ecological rationality" for this concept.

8.5 Summary

The basic ideas from this chapter can be summarized as follows:

1. Indifference zone approaches require the specification of the distance δ^* and the probability of a correct selection P^*. The indifference zone procedure (Fig. 8.1) assumes known and common variances.
2. Subset selection requires the specification of the number of samples r and the probability of a correct selection P^*.
3. Threshold selection requires the specification of the number of samples r and the probability of a correct selection P^*.
4. The probability of a correct acceptance P^* in TS is related to the error of the first kind in hypothesis testing.
5. Factorial designs are useful to detect interactions.
6. Implementation details can have a significant influence on the algorithm's performance.
7. An annealing schedule can be used to adapt the probability of a correct acceptance during the search process of an optimization algorithm.
8. Threshold selection can be interpreted as a rule of inductive behavior, or as an automatic testing rule.
9. Selection schemes require the specification of scientifically important differences. The determination of these values lies outside the domain of statistics.
10. The 1/5 rule has to be modified if the function values are disturbed by noise.
11. Threshold selection can be characterized as a fast and frugal heuristic for decision making under uncertainty. Obviously, TS does not work in every environment.

8.6 Further Reading

Bechhofer et al. (1995) give an in-depth presentation of methods for statistical selection, screening, and comparisons. Branke et al. (2005) compare ranking and selection methods. Schwefel (1995), Rudolph (1997a), and Beyer (2001) present further analyses of evolution strategies. Arnold & Beyer (2003) and Jin & Branke (2005) discuss optimization algorithms in noisy environments. Rubinstein (1998) and Gigerenzer & Selten (2002) introduce models of bounded rationality. The concept of simple heuristics is presented in (Gigerenzer et al. 1999).

9

Summary and Outlook

> It is good to have an end to journey toward;
> but it is the journey that matters, in the end.
> —Ernest Hemingway

Now that we have reached the end of an exploratory tour during which we discussed a broad spectrum of ideas from computer science, philosophy of science, and statistics, it is time to summarize the basic achievements. To compare different objects is a basic human activity. Decisions are based on comparisons. However, the accuracy of the observed data that are necessary for comparisons is limited in many real-world situations.

Statistics provides a means to cope with this uncertainty or "noise". Although computer science is built on deterministic grounds, it can be advantageous to introduce uncertainty or randomness. One famous example is the quick-sort algorithm, where randomness is introduced to implement the selection procedure. Another important field is stochastic search algorithms which started their triumphal procession in the 1960s. Nowadays, stochastic search algorithms belong to the standard repertoire of every optimization practitioner who has to solve harder problems than just "toy" problems. However, their enormous flexibility complicates their analysis. The approach presented in this book suggests treating program runs as experiments to enable a statistical analysis.

9.1 The New Experimentalists

Experiments have a long history in science, but their role changed drastically over recent centuries. They have been downplayed by Aristotelians, who favored deduction from first principles for a long time. But the scientific revolution of the seventeenth century declared the experimental method as the "royal road to knowledge." The first scientific journals that presented experimental results and deductions from experiments were established at that time—a completely different situation to the theoretically oriented contents of modern scientific journals. Hacking (1983), one of the most influential contemporary philosophers of science, proposes to "initiate a Back-to-Bacon movement, in which we attend more seriously to experimental science." His slogan "an experiment may have a life of its own" points to several new experimentalist themes, but it does not claim that experimental work could exist independently of theory. "That would be the blind work of those whom Bacon mocked as 'mere empirics'. It remains the case, however, that much truly fundamental research precedes any relevant theory whatsoever" (Hacking 1983).

Ian Hacking, Robert Ackermann (1989), Nancy Cartwright (1983, 2000), Allen Franklin (1990), Peter Galison (1987), Ronald Giere (1999), and Deborah Mayo (1996) belong to a group of philosophers of science who share the thesis that "focusing on aspects of experiments holds the key to avoiding or solving a number of problems, problems thought to stem from the tendency to view science from theory-dominated stances." Ackermann (1989) introduced the term "new experimentalists." One major goal of the new experimentalists is to develop statistical tools for generating reliable data from experiments and using such data to learn from experiments. Mayo (1996) proposes a modern theory of statistical testing and learning from error. This book is an attempt to establish a modern theory of statistical testing in computer science, especially in evolutionary computation. The *new experimentalism in evolutionary computation* has its roots in the actual debate over the epistemology of experimentation in philosophy. Based on the ideas presented by the new experimentalists, in particular on Mayo's *learning from error* and her concept of *severity*, a methodology for performing and analyzing computer experiments has been developed. By controlling the errors that occur during experimentation, we can gain insight into the dependencies and interactions of important factors.

The new experimentalists extend Popper's position that only hypotheses that are in principle falsifiable by experience should count as scientific. The resulting consequences from this position have been widely discussed in recent decades, so we mention only one problem that arises from the Popperian view: Hypotheses require assumptions. A serious problem arises when we have to decide whether the hypothesis itself or the supporting assumption is wrong. Moreover, these assumptions require additional assumptions, which leads to an infinite regress.

9.2 Learning from Error

This book discusses various ways to pose the right questions, to measure the performance of algorithms, and to analyze the results. However, the statistical analysis is only the first part of the investigation, that is, it is the beginning, and not the end. We learn about algorithms by being perspicacious investigators knowing how to produce errors. Actively generating errors is a major step forward in understanding how algorithms work. Various sources of error in the context of evolutionary computation are pointed out. Errors can be caused by the selection of an inadequate test function, erroneously specified experimental designs, wrongly specified experimental goals, inadequately selected performance measures, and misinterpretations of the experimental or the statistical results. These errors, which were mentioned in Chap. 1, are revisited now. Means to master Problems 1.1 to 1.4 and additional ones that occurred during our analyses are:

Answer (to Problem 1.1). *The lack of standardized test functions, or benchmark problems.* Problems related to test suites were discussed in Chap. 4. We developed an elevator simulation model (S-ring) that can be used to generate test problem instances. The results are of practical relevance. They are based on an intensive cooperation with one of the world's leading elevator manufacturers (Markon et al. 2001; Beielstein & Markon 2002; Beielstein et al. 2003a, b; Bartz-Beielstein et al. 2003c, 2005c; Bartz-Beielstein & Markon 2004). A complete book is devoted to the S-ring model and related optimization problems (Markon et al. 2006).

Suganthan et al. (2005) accentuate the need to evaluate test functions "in a more systematic manner by specifying a common termination criterion, size of problems, initialization scheme, linkages/rotation, etc. There is also a need to perform a scalability study demonstrating how the running time/evaluations increase with an increase in the problem size." They established a special session on real-parameter optimization during CEC 2005. A new standard test suite which includes some real world problems was proposed.

The development of realistic test suites is necessary, but only the first step. It should be complemented with the development of statistical tools, e.g., tests as learning tools (cf. Chap. 2).

Answer (to Problem 1.2). *The usage of different (or inadequately selected) performance measures.* We distinguish different performance measures to analyze algorithms, e.g., efficiency and effectivity. This classification is based on ideas that Schwefel (1977) presented nearly three decades ago. Nowadays it is a well-accepted fact that there is no computer algorithm that performs better than any other algorithm in all cases (Droste et al. 2000). However, the interaction between the problem (environment, resources) and the algorithm is crucial for its performance. To demonstrate an effect, a test problem that is well-suited to the solver (algorithm) must be chosen. *Ceiling effects* can make results from computer experiments useless. They occur when every algorithm achieves the maximum level of performance—the results are indistinguishable. *Floor effects* arise when the problem is too hard, so no algorithm can produce a satisfactory solution and every statistical analysis will detect no difference in the performance. Statistical methods such as run-length distributions have been proposed to tackle this issue. This problem was addressed in Chap. 7.

Answer (to Problem 1.3). *The impreciseness of results, and therefore no clearly specified conclusions.* To prevent misinterpretations of the experimental results, guidelines from experimental algorithmics are recommended. These guidelines, e.g., to state a clear set of objectives, or to formulate a question or a hypothesis, have been extended and reformulated from the perspective of an error statistician in this work. Analysis of variance methods as well as regression methods and hypothesis testing were presented in Chaps. 3 and 7.

Answer (to Problem 1.4). *The lack of reproducibility of experiments.* Erroneously specified experimental designs, e.g., wrongly selected exogenous strategy parameters or problem parameters, can cause this problem. *Designs* play a key role in this context: The concept of problem and algorithm designs is consequently realized. Experimental designs are used to vary and control these errors systematically. The experimenter can screen out less important factors and concentrate the analysis on the relevant ones. To give an example: Evolutionary algorithms produce random results. The experimenter can vary the input parameters of the algorithm, e.g., change the recombination operator. This leads directly to the central question "How much variance in the response is explained by the variation in the algorithm?" Statistical tools that are based on the analysis of variance methodology can be used to tackle this question. Classical ANOVA and modern regression techniques like tree-based regression or design and analysis of computer experiments (DACE) follow this principle. Another basic tool to perform a statistical analysis is *hypothesis testing.*

We are not the first to use design of experiments techniques to analyze algorithms. However, the first attempts to apply design of experiments techniques to evolution strategies and particle swarm optimization have been presented in Beielstein et al. (2001) and Beielstein & Markon (2001). We have developed the sequential parameter optimization method: SPO combines methods from classical DOE, computational statistics, and design and analysis of computer experiments. Results from the SPO can be analyzed with NPT* tools: The experimenter can learn from errors while improving an algorithm, see also Problems 9.1 and 9.2. We consider the NPT* analysis as the crucial step in the analysis of computer algorithms. Experimental designs have been introduced in Chap. 5.

In addition to these problems listed by Eiben & Jelasity (2002), the following problems are of importance:

Problem 9.1. *Wrongly specified experimental goals.* Gary Klein (2002) uses the term "fiction of optimization" to characterize this problem.

Answer. Boundary conditions that are necessary to perform optimization tasks have been discussed in Sect. 7.1. Specifying and analyzing boundary conditions is in accordance with Mayo's concept of learning from error and one important step of the SPO approach. Experimental goals were discussed in Chap. 7.

Problem 9.2. *Misinterpretations of the statistical results.* Serious problems arise when statistical significance and scientific meaning are not distinguished.

Answer. Introducing different models provides statistical tools to deal with this problem. Based on NPT*, Mayo's extension of the classical Neyman–Pearson theory of statistical testing, we developed statistical tools that allow the objective comparison of experimental results. Misconstruals can occur if statistical tests are not severe. Consider, for example, the first misconstrual (MC-1) from Sect. 2.5 that can be accomplished by increasing the sample size n or by reducing the significance level:

> A test can be specified that will produce a result that exceeds a prespecified difference by the required difference. As a consequence, the null hypothesis H is rejected, even if the true difference exceeds the prespecified difference by as little as one likes.

We developed plots of the observed significance level (OSL) as key elements for an extended understanding of the significance of statistical results. They are easy to interpret and combine information about the p-value, the sample size, and the experimental error. A bootstrap procedure to generate OSL plots independently from any assumptions on the underlying distribution was introduced. Misinterpretations of the statistical results were discussed in Chaps. 3 and 8.

9.3 Theory and Experiment

This book does not solely transfer concepts to compare and improve algorithms from statistics to computer science. It presents a self-contained experimental methodology that bridges the gap between theory and experiment. The advantage of applying results from theory, for example, Beyer (2001), to real-world optimization problems can be analyzed in an objective manner. However, as a consequence of our considerations, the interpretation of the scientific import of these results requires human experience, or the "experimenter's skill."

Why is the experimenter's skill central in our argumentation? The experimenter's skill comprises the ability to get the apparatus to indicate phenomena in a certain way. Numerous examples from the history of science can be listed in which the invention of a new apparatus enables the experimenter to perform another investigation. Results from these experiments defined the route that the theoreticians must follow. Gigerenzer's (2003) tool-to-theory approach extends this idea from technical apparatus to abstract entities such as statistical procedures. We summarize an example presented in Hacking (1983) to illustrate our argumentation.

Example 9.1 (The Faraday Effect). The Faraday effect, or magneto-optical effect, describes the rotation of the plane of polarization (plane of vibration) of a light beam by a magnetic field (Encyclopaedia Britannica Online 2001). Being a deeply religious man, Michael Faraday (1791–1867) was convinced that all forces in nature must be connected. At that time the Newtonian unity of science was in confusion due to several important discoveries, i.e., the wave theory of light. Faraday unsuccessfully tried to establish a connection between electrification and light in 1822, in 1834, and in 1844. In 1845 he gave up and tried to discover a connection between the forces of electromagnetism and light. Using a special kind of dense glass, which had been developed earlier in a different context, he discovered the magneto-optical effect. Faraday had no theory of what he found. One year later, G.B. Airy integrated the experimental observations into the wave theory of light simply by adding some ad hoc further terms to the corresponding equations. "This is a standard move in physics. In order to make the equations fit the phenomena, you pull from the shelf some fairly standard extra terms for the equations, without knowing why one rather than another will do the trick." Only 47 years later, in 1892, H.A. Lorentz combined models proposed by Kelvin and adapted by Maxwell with his electron theory. ■

This example nicely illustrates several levels of theory. Theory, as mentioned earlier, can be characterized as *speculation*: It can be seen as the process of restructuring thoughts or playing with ideas that are based on a qualitative understanding of some general features from reality. However, there is no direct link between theory and experiment. Most initial thoughts are not directly testable. Here *calculation* comes into play. Calculation is the mathematical formulation to bring speculative thoughts into accordance with the world and to conduct an experimental verification. Calculation is the first part to bridge the gap between theory and experiment.

We have not left the classical, hypothetico-deductive grounds so far. However, to bring theory in accordance with reality is not simply a matter of calculation. To do this requires more than just quantifying speculative thoughts. The idea of beginning with speculations that are gradually cast into a form from whence experimental tests can be deduced, appears to be attractive—but it is incomplete. A very extensive activity is necessary: model building. Models can be all sorts of things (recall the discussion in Sect. 2.4). It is crucial for our reasoning that they can coexist in theory. Despite the common understanding that at most one model can be true, several models of the physical world can be used indifferently and interchangeably in the theoretical context. Hacking presents a typical sentence from a physics textbook as an illustrative example:

> For free particles, however, we may take either the advanced or retarded potentials, or we may put the results in a symmetrical form, without affecting the result (Mott & Sneddon 1948).

Hence, models are not merely intermediaries that connect some abstract aspects of real phenomena by simplifying mathematical structures to theories that govern the phenomena. Why can physicists use a number of mutually inconsistent models within the same theory? Recalling the ideas presented in Chap. 1, we can state that models are the central elements of science. Models are more robust than theory, that is, "you keep the model and dump the theory." The number of models scientists use in their daily routine increases from year to year. Maybe there will be one unified theory of all in some years—but "that will leave most physics intact, for we shall have to do applied physics, working out what happens from case to case" (Hawking 1980).

Approximations appear to be a solution to bridge the gap between models for theory and models for reality. But the number of possible approximations is endless, and the correct approximation cannot be derived from theory. Going one step further, Nancy Cartwright claims that "theory itself has no truth in it" (Cartwright 1983, 2000). We follow Hacking (1983), who gives a descriptive characterization of the interplay between theories, models, and reality:

> I myself prefer an Argentine fantasy. God did not write a Book of Nature of the sort that the old Europeans imagined. He wrote a Borgesian library, each book of which is as brief as possible, yet each book of which is inconsistent with each other. No book is redundant. For every book, there is some humanly accessible bit of Nature such that that book, and no other, makes possible the comprehension, prediction and influencing what is going on. Far from being untidy, this is the New World Leibnizianism. Leibniz said that God chose a world which maximized the variety of phenomena while choosing the simplest laws. Exactly so: but the best way to maximize phenomena and have the simplest laws is to have the laws inconsistent with each other, each applying to this or that but none applying to all.

The methodology presented in this book may be the missing link needed by the practitioner to consciously apply theoretical results to practical problems—and by the theoretician to explore new ideas and to confront speculations with reality. Figure 9.1 illustrates a modified view of the relationship between theory and experiment from Chap. 1.

9.4 Outlook

The items discussed in this book suggest various routes for further research. We will list some of them.

The experimental approach presented in this work may lay the cornerstone for a "Borgesian library" in evolutionary computation. Consider a theory T_1, e.g., entitled "Evolutionary Algorithms in Theory," and a theory T_2, entitled "The Theory of Evolution Strategies" (Fig. 9.1). Both theories use models as tools for representing parts of the world (or the theory) for specific purposes. We distinguished representational models, which represent a certain part of the world, from instantial models, that are used to present abstract entities. Performing experiments, data can be generated to test the fit of the model with some part of the world or with some theory. As models are limited per definitionem, they may contain laws that are inconsistent with each other.

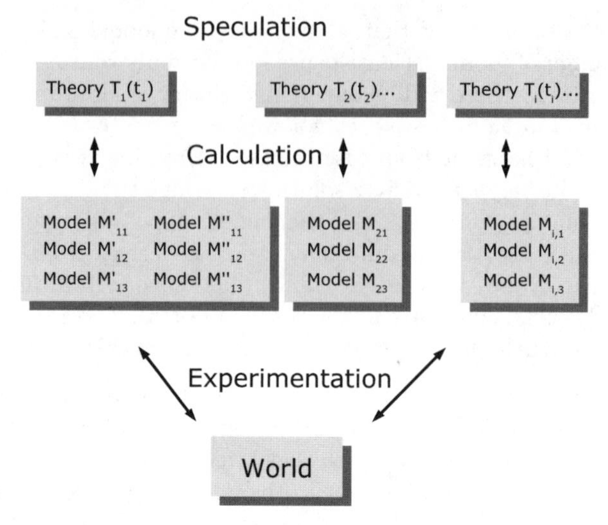

Fig. 9.1. A second attempt to model the relationship between theory and practice. The first attempt (Fig. 2.1) is reconsidered. Different theories and models, even with conflicting laws, coexist. The variables t_i denote the time-dependency of some theories. Speculation can be interpreted as "playing with ideas," calculation brings speculative thoughts in accordance with models, and experimentation tests the fit of models with the world

And, not only models for different theories may contain conflicting laws—even models that are used within one theory might lead to different conclusions. At this point the approach presented in this book becomes relevant: The experimenter can use statistical tools to investigate the error probabilities by "actively probing, manipulating, and simulating patterns of error, and by deliberately introducing known patterns of error into the collection and analysis of data"(Mayo 1996). Consider two scenarios:

1. Experimental designs (Chap. 5) provide means to specify the essential conditions in an objective manner. Optimization practitioners can consult a "book from the problem design section of the library" and look up a candidate algorithm that might be able to solve their problem. This algorithm will not be used directly with some default parameter settings—it will be tuned before the optimization run is performed.
2. Researchers will base the comparison of different algorithms not on their default parameterizations, but on the tuned versions. The SPO (or a similar method) enables an algorithmical tuning process with traceable costs.

Consequently, a *tool box* with many different algorithms, e.g., as suggested by Schwefel (1995), "might always be the 'optimum optimorum' for the practitioner." The methods presented in this work might give some valuable advice for the selection of an appropriate tool.

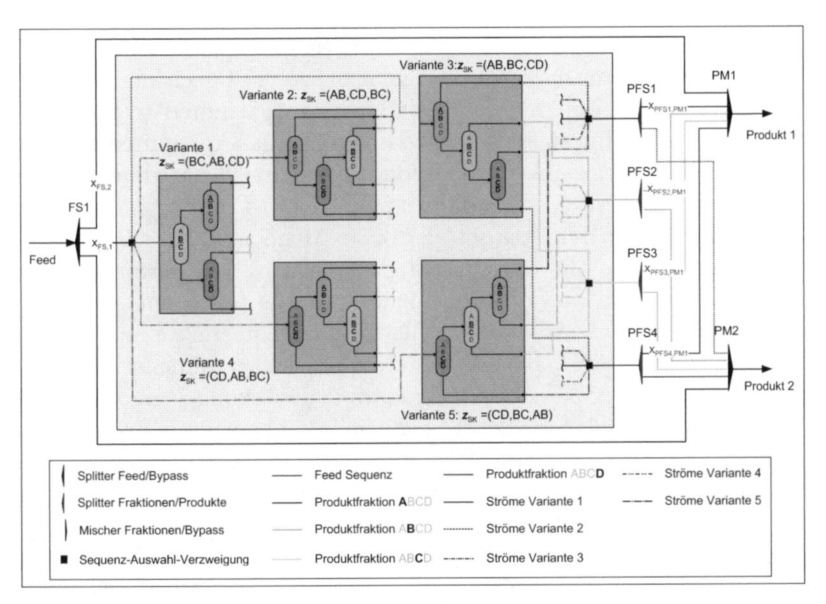

Fig. 9.2. An evolution strategy applied to optimize the design of a nonsharp separation sequence. The German words "Sequenz-Auswahl-Verzweigung," "Ströme Variante," and "Mischer" can be translated to English as "sequence-branching point," "stream variant," and "mixer," respectively. SPO could significantly improve the performance of the ES. Source: Frank Henrich, RWTH Aachen, private communication

Recent discussions indicated a great demand for an *automated version* of the sequential parameter optimization procedure (Chap. 7). The development of such an automated tool is—at least from our perspective—a conflicting goal, because the user does not "see" what happens during the design optimization. However, SPO will gain acceptance and influence, if we keep the complexity that is necessary to apply SPO as low as possible. A first implementation of an automated SPO is under development (Bartz-Beielstein et al. 2005b).

SPO has been proven useful in other problem domains. We mention one result from the chemical engineering domain (Fig. 9.2): An evolution strategy was applied to optimize the design of a nonsharp separation sequence (Preuß & Bartz-Beielstein 2006). SPO was applied successfully to improve the performance of the ES. This problem and the appropriate simulation techniques were developed and analyzed at the Chair of Technical Thermodynamics of the Rheinisch-Westfälischen Technischen Hochschule Aachen (RWTH Aachen University of Technology, Germany). SPO has also been applied by Tosic (2006) to optimize the runtime parameters of a genetic algorithm that planarizes a given graph with the aim to minimize the number of crossings in its drawing.

Including the relationship between step-size adaptation and threshold selection into the analysis from the case study in Chap. 8 will provide interesting insights into the behavior of evolution strategies. In general, an analysis of the *self-adaptation* mechanisms seems to be one of the most exciting tasks for further research. Recall that a careful examination is required to perform this task, because the evolution strategy presented in Chap. 6 required the specification of nine design variables and various interactions have to be considered.

First attempts have been made to apply SPO to *multicriteria optimization* problems (Bartz-Beielstein et al. 2003b; Mehnen et al. 2004a, 2005). A discussion—similar to the one presented in Chap. 7—of performance measures for multicriteria optimization algorithms has to be done in advance. The S-metric selection evolutionary algorithm (Emmerich et al. 2005) was analyzed with SPO. Of particular interest are higher dimensional solution spaces and fitness approximation techniques in evolutionary algorithms (Emmerich & Naujoks 2004; Naujoks et al. 2005b, a). These issues are analyzed in the collaborative research center "Design and Management of Complex Technical Processes and Systems by Means of Computational Intelligence Methods" (Beielstein et al. 2003c).

Visual tools that enable an intuitive understanding of experimental results and their scientific meaning should be developed. The observed significance level plots (Chap. 1) are merely a first step in this direction.

Based on considerations related to the concept of *bounded rationality*, one can ask in which environment an algorithm performs well (Gigerenzer et al. 1999). Not merely organic evolution, but also social evolution might give valuable hints to develop new strategies or to understand existing behaviors. This process can be beneficial in both directions: Evolutionary algorithms can be used to model social behavior, and vice versa. Consider, for example, the model of urban growth by cellular automata from Bäck et al. (1996), or a current diploma thesis that models farm size and market power on agricultural land markets with particle swarm optimization (de Vegt 2005).

We close the final chapter of this book with an analogy to depict the relativity of good algorithms (or strategies) and to demonstrate the usefulness of experience:

> In a remote stream in Alaska, a rainbow trout spies a colorful dimple on the undersurface of the water with an insect resting on top of it. Darting over with the mouth agape, the fish bites down and turns in search for its next victim. It does not get far, however, before the "insect" strikes back. The trout is yanked from the quiet stream by the whiplike pull of a fly fisherman's rod. In a world without fisherman, striking all the glitter is adaptive; it increases the chance for survival. In a world with predators, however, this once-adaptive strategy can turn a feeding fish into a fisherman's food (Goldstein et al. 2002).

References

Ackermann, R. (1989). The new experimentalism. *British Journal for the Philosophy of Science*, 40, 185–190.

Aggarwal, A. & Floudas, C. A. (1990). Synthesis of general distillation sequences—nonsharp separations. *Computers & Chemical Engineering*, 14, 631–653.

Aho, A., Johnson, D., Karp, R., Kosaraju, S., et al. (1997). Merging opportunities for theoretical computer science. *SIGACT News*, 28(3), 65–74.

Anderson, R. (1997). The role of experiment in the theory of algorithms. In *Proceedings of the 5th DIMACS Challenge Workshop*, volume 59 of *DIMACS: Series in Discrete Mathematics and Theoretical Computer Science* (pp. 191–196). Providence RI: American Mathematical Society.

Arnold, D. V. (2001). Evolution strategies in noisy environments—a survey of existing work. In L. Kallel, B. Naudts, & A. Rogers (Eds.), *Theoretical Aspects of Evolutionary Computing* (pp. 239–249). Berlin, Heidelberg, New York: Springer.

Arnold, D. V. & Beyer, H.-G. (2003). A comparison of evolution strategies with other direct search methods in the presence of noise. *Computational Optimization and Applications*, 24(1), 135–159.

Aslett, R., Buck, R. J., Duvall, S. G., Sacks, J., & Welch, W. J. (1998). Circuit optimization via sequential computer experiments: design of an output buffer. *Journal of the Royal Statistical Society: Series C (Applied Statistics)*, 47(1), 31–48.

Athen, H. & Bruhn, J., Eds. (1980). *Lexikon der Schulmathematik. Studienausgabe.* Köln, Germany: Aulis.

Azadivar, F. (1999). Simulation optimization methodologies. In *WSC '99: Proceedings of the 31st Winter Simulation Conference* (pp. 93–100). New York NY: Association for Computing Machinery.

Bäck, T. (1996). *Evolutionary Algorithms in Theory and Practice.* New York NY: Oxford University Press.

Bäck, T., Beielstein, T., Naujoks, B., & Heistermann, J. (1995). Evolutionary algorithms for the optimization of simulation models using PVM. In J. Dongarra, M. Gengler, B. Tourancheau, & X. Vigouroux (Eds.), *Second European PVM Users' Group Meeting (EuroPVM'95)* (pp. 277–282). Paris, France: Hermès.

Bäck, T., Dörnemann, H., Hammel, U., & Frankhauser, P. (1996). Modeling urban growth by cellular automata. In H.-M. Voigt, W. Ebeling, I. Rechenberg, & H.-P.

Schwefel (Eds.), *Proceedings Parallel Problem Solving from Nature—PPSN IV, Berlin* (pp. 636–645). Berlin, Heidelberg, New York: Springer.

Bandler, J., Cheng, Q., Dakroury, S., Mohamed, A., Bakr, M., Madsen, K., & Søndergaard, J. (2004). Space mapping: the state of the art. *IEEE Transactions on Microwave Theory and Techniques*, 52(1), 337–361.

Banks, J., Carson, J. S., Nelson, B. L., & Nicol, D. M. (2001). *Discrete Event System Simulation.* Upper Saddle River NJ: Prentice Hall.

Barney, G. (1986). *Elevator Traffic Analysis, Design and Control.* Cambridge, U.K.: Cambridge University Press.

Barr, R., Golden, B., Kelly, J., Rescende, M., & Stewart, W. (1995). Designing and reporting on computational experiments with heuristic methods. *Journal of Heuristics*, 1(1), 9–32.

Barr, R. & Hickman, B. (1993). Reporting computational experiments with parallel algorithms: Issues, measures, and experts' opinions. *ORSA Journal on Computing*, 5(1), 2–18.

Bartz-Beielstein, T. (2003). *Experimental Analysis of Evolution Strategies—Overview and Comprehensive Introduction.* Interner Bericht des Sonderforschungsbereichs 531 Computational Intelligence CI–157/03, Universität Dortmund, Germany.

Bartz-Beielstein, T. (2005a). Evolution strategies and threshold selection. In M. J. Blesa Aguilera, C. Blum, A. Roli, & M. Sampels (Eds.), *Proceedings Second International Workshop Hybrid Metaheuristics (HM'05)*, volume 3636 of *Lecture Notes in Computer Science* (pp. 104–115). Berlin, Heidelberg, New York: Springer.

Bartz-Beielstein, T. (2005b). *New Experimentalism Applied to Evolutionary Computation.* PhD thesis, Universität Dortmund, Germany.

Bartz-Beielstein, T., Blum, D., & Branke, J. (2005a). Particle swarm optimization and sequential sampling in noisy environments. In R. Hartl & K. Doerner (Eds.), *Proceedings 6th Metaheuristics International Conference (MIC2005)* (pp. 89–94). Vienna, Austria.

Bartz-Beielstein, T., de Vegt, M., Parsopoulos, K. E., & Vrahatis, M. N. (2004a). *Designing Particle Swarm Optimization with Regression Trees.* Interner Bericht des Sonderforschungsbereichs 531 Computational Intelligence CI–173/04, Universität Dortmund, Germany.

Bartz-Beielstein, T., Lasarczyk, C., & Preuß, M. (2005b). Sequential parameter optimization. In B. McKay & others (Eds.), *Proceedings 2005 Congress on Evolutionary Computation (CEC'05), Edinburgh, Scotland*, volume 1 (pp. 773–780). Piscataway NJ: IEEE Press.

Bartz-Beielstein, T., Limbourg, P., Mehnen, J., Schmitt, K., Parsopoulos, K. E., & Vrahatis, M. N. (2003a). Particle swarm optimizers for pareto optimization with enhanced archiving techniques. In R. Sarker & others (Eds.), *Proceedings 2003 Congress on Evolutionary Computation (CEC'03), Canberra*, volume 3 (pp. 1780–1787). Piscataway NJ: IEEE.

Bartz-Beielstein, T., Limbourg, P., Mehnen, J., Schmitt, K., Parsopoulos, K. E., & Vrahatis, M. N. (2003b). *Particle Swarm Optimizers for Pareto Optimization with Enhanced Archiving Techniques.* Interner Bericht des Sonderforschungsbereichs 531 Computational Intelligence CI–153/03, Universität Dortmund, Germany.

Bartz-Beielstein, T. & Markon, S. (2004). Tuning search algorithms for real-world applications: A regression tree based approach. In G. W. Greenwood (Ed.), *Proceedings 2004 Congress on Evolutionary Computation (CEC'04), Portland OR*, volume 1 (pp. 1111–1118). Piscataway NJ: IEEE.

Bartz-Beielstein, T., Markon, S., & Preuß, M. (2003c). Algorithm based validation of a simplified elevator group controller model. In T. Ibaraki (Ed.), *Proceedings 5th Metaheuristics International Conference (MIC'03)* (pp. 06/1–06/13 (CD–ROM)). Kyoto, Japan.

Bartz-Beielstein, T. & Naujoks, B. (2004). *Tuning Multicriteria Evolutionary Algorithms for Airfoil Design Optimization.* Interner Bericht des Sonderforschungsbereichs 531 Computational Intelligence CI–159/04, Universität Dortmund, Germany.

Bartz-Beielstein, T., Parsopoulos, K. E., & Vrahatis, M. N. (2004b). Analysis of particle swarm optimization using computational statistics. In T.-E. Simos & C. Tsitouras (Eds.), *Proceedings International Conference Numerical Analysis and Applied Mathematics (ICNAAM)* (pp. 34–37). Weinheim, Germany: Wiley-VCH.

Bartz-Beielstein, T. & Preuß, M. (2004). Experimental research in evolutionary computation (tutorial). Congress on Evolutionary Computation (CEC 2004), Portland OR. `http://ls11-www.cs.uni-dortmund.de/people/tom`. Cited 30 June 2004.

Bartz-Beielstein, T. & Preuß, M. (2005a). Experimental research in evolutionary computation (tutorial). Genetic and Evolutionary Computation Conf. (GECCO 2005), Washington DC. `http://ls11-www.cs.uni-dortmund.de/people/tom`. Cited 30 June 2005.

Bartz-Beielstein, T. & Preuß, M. (2005b). Experimental research in evolutionary computation (tutorial). Congress on Evolutionary Computation (CEC 2005), Edinburgh UK. `http://ls11-www.cs.uni-dortmund.de/people/tom`. Cited 10 October 2004.

Bartz-Beielstein, T., Preuß, M., & Markon, S. (2005c). Validation and optimization of an elevator simulation model with modern search heuristics. In T. Ibaraki, K. Nonobe, & M. Yagiura (Eds.), *Metaheuristics: Progress as Real Problem Solvers*, Operations Research/Computer Science Interfaces (pp. 109–128). Berlin, Heidelberg, New York: Springer.

Bartz-Beielstein, T., Preuß, M., & Reinholz, A. (2003d). Evolutionary algorithms for optimization practitioners (tutorial). Proceedings 5th Metaheuristics International Conference (MIC'03) Kyoto, Japan. `http://ls11-www.cs.uni-dortmund.de/people/tom`. Cited 3 September 2003.

Bartz-Beielstein, T., Schmitt, K., Mehnen, J., Naujoks, B., & Zibold, D. (2004c). *KEA—A Software Package for Development, Analysis, and Application of Multiple Objective Evolutionary Algorithms.* Interner Bericht des Sonderforschungsbereichs 531 Computational Intelligence CI–185/04, Universität Dortmund, Germany.

Bechhofer, R. E., Dunnett, C. W., Goldsman, D. M., & Hartmann, M. (1990). A comparison of the performances of procedures for selecting the normal population having the largest mean when populations have a common unknown variance. *Communications in Statistics*, B19, 971–1006.

Bechhofer, R. E., Santner, T. J., & Goldsman, D. M. (1995). *Design and Analysis of Experiments for Statistical Selection, Screening, and Multiple Comparisons.* New York NY: Wiley.

Beielstein, T. (2003). *Tuning Evolutionary Algorithms—Overview and Comprehensive Introduction.* Interner Bericht des Sonderforschungsbereichs 531 *Computational Intelligence* CI–148/03, Universität Dortmund, Germany.

Beielstein, T., Dienstuhl, J., Feist, C., & Pompl, M. (2001). *Circuit Design Using Evolutionary Algorithms.* Interner Bericht des Sonderforschungsbereichs 531 *Computational Intelligence* CI–122/01, Universität Dortmund, Germany.

Beielstein, T., Dienstuhl, J., Feist, C., & Pompl, M. (2002a). Circuit design using evolutionary algorithms. In D. B. Fogel & others (Eds.), *Proceedings 2002 Congress on Evolutionary Computation (CEC'02) Within Third IEEE World Congress on Computational Intelligence (WCCI'02), Honolulu HI* (pp. 1904–1909). Piscataway NJ: IEEE.

Beielstein, T., Ewald, C.-P., & Markon, S. (2003a). Optimal elevator group control by evolution strategies. In E. Cantú-Paz & others (Eds.), *Proceedings Genetic and Evolutionary Computation Conf. (GECCO 2003), Chicago IL, Part II*, volume 2724 of *Lecture Notes in Computer Science* (pp. 1963–1974). Berlin, Heidelberg, New York: Springer.

Beielstein, T. & Markon, S. (2001). *Threshold Selection, Hypothesis Tests, and DOE Methods.* Interner Bericht des Sonderforschungsbereichs 531 Computational Intelligence CI–121/01, Universität Dortmund, Germany.

Beielstein, T. & Markon, S. (2002). Threshold selection, hypothesis tests, and DOE methods. In D. B. Fogel & others (Eds.), *Proceedings 2002 Congress on Evolutionary Computation (CEC'02) Within Third IEEE World Congress on Computational Intelligence (WCCI'02), Honolulu HI* (pp. 777–782). Piscataway NJ: IEEE.

Beielstein, T., Markon, S., & Preuß, M. (2003b). A parallel approach to elevator optimization based on soft computing. In T. Ibaraki (Ed.), *Proceedings 5th Metaheuristics International Conference (MIC'03)* (pp. 07/1–07/11 (CD–ROM)). Kyoto, Japan.

Beielstein, T., Mehnen, J., Schönemann, L., Schwefel, H.-P., Surmann, T., Weinert, K., & Wiesmann, D. (2003c). Design of evolutionary algorithms and applications in surface reconstruction. In H.-P. Schwefel, I. Wegener, & K. Weinert (Eds.), *Advances in Computational Intelligence—Theory and Practice* (pp. 145–193). Berlin, Heidelberg, New York: Springer.

Beielstein, T., Parsopoulos, K. E., & Vrahatis, M. N. (2002b). *Tuning PSO parameters through sensitivity analysis.* Interner Bericht des Sonderforschungsbereichs 531 *Computational Intelligence* CI–124/02, Universität Dortmund, Germany.

Belisle, C. J. P. (1992). Convergence theorems for a class of simulated annealing algorithms. *Journal Applied Probability*, 29, 885–895.

Bentley, P. (2002). ISGEC workshop on standards at GECCO 2002. `http://www.cs.ucl.ac.uk/staff/P.Bentley/standards.html`. Cited 6 April 2004.

Beyer, H.-G. (2000). Evolutionary algorithms in noisy environments: Theoretical issues and guidelines for practice. *CMAME (Computer Methods in Applied Mechanics and Engineering)*, 186, 239–267.

Beyer, H.-G. (2001). *The Theory of Evolution Strategies.* Berlin, Heidelberg, New York: Springer.

Beyer, H.-G. & Schwefel, H.-P. (2002). Evolution strategies—A comprehensive introduction. *Natural Computing*, 1, 3–52.

Birattari, M., Stützle, T., Paquete, L., & Varrentrapp, K. (2002). A racing algorithm for configuring metaheuristics. In W. Langdon (Ed.), *GECCO 2002: Proceedings of the Genetic and Evolutionary Computation Conference* (pp. 11–18). San Francisco CA: Morgan Kaufmann.

Box, G. E. P. (1957). Evolutionary operation: A method for increasing industrial productivity. *Applied Statistics*, 6, 81–101.

Box, G. E. P. & Draper, N. R. (1987). *Empirical Model Building and Response Surfaces.* New York NY: Wiley.

Box, G. E. P., Hunter, W. G., & Hunter, J. S. (1978). *Statistics for Experimenters.* New York NY: Wiley.

Branke, J., Chick, S., & Schmidt, C. (2005). New developments in ranking and selection: An empirical comparison of the three main approaches. In M. E. Kuhl & others (Eds.), *Proceedings of the 2005 Winter Simulation Conference* (pp. 708–717). Piscataway NJ: IEEE.

Branke, J., Schmidt, C., & Schmeck, H. (2001). Efficient fitness estimation in noisy environments. In L. Spector (Ed.), *Genetic and Evolutionary Computation Conference (GECCO'01)* (pp. 243–250). San Francisco CA: Morgan Kaufmann.

Breiman, L., Friedman, J. H., Olshen, R. A., & Stone, C. J. (1984). *Classification and Regression Trees.* Monterey CA: Wadsworth.

Briest, P., Brockhoff, D., Degener, B., et al. (2004). Experimental supplements to the theoretical analysis of EAs on problems from combinatorial optimization. In X. Yao, E. Burke, J. A. Lozano, & others (Eds.), *Parallel Problem Solving from Nature—PPSN VIII*, volume 3242 of *Lecture Notes in Computer Science* (pp. 21–30). Berlin, Heidelberg, New York: Springer.

Broyden, C. G. (1970). The convergence of a class of double-rank minimization algorithms. *Journal of the Institute of Mathematics and Its Applications*, 6, 76–90.

Bussieck, M., Drud, A., Meeraus, A., & Pruessner, A. (2003). Quality assurance and global optimization. In *Global Optimization and Constraint Satisfaction: First International Workshop on Global Constraint Optimization and Constraint Satisfaction, COCOS 2002*, volume 2861 of *Lecture Notes in Computer Science* (pp. 223–238). Berlin, Heidelberg, New York: Springer.

Cartwright, N. (1983). *How the Laws of Physics Lie.* Oxford, U.K.: Oxford University Press.

Cartwright, N. (2000). *The Dappled World: A Study of the Boundaries of Science.* Cambridge, U.K.: Cambridge University Press.

Chalmers, A. F. (1999). *What Is This Thing Called Science.* St. Lucia, Australia: University of Queensland Press.

Chambers, J., Cleveland, W., Kleiner, B., & Tukey, P. (1983). *Graphical Methods for Data Analysis.* Belmont CA: Wadsworth.

Chambers, J. M. & Hastie, T. H., Eds. (1992). *Statistical Models in S.* Pacific Grove CA: Wadsworth & Brooks/Cole.

Chen, H. C., Chen, C. H., Dai, L., & Yücesan, E. (1997). New development of optimal computing budget allocation for discrete event simulation. In S. Andradóttir, K. J. Healy, D. H. Withers, & B. L. Nelson (Eds.), *Proceedings of the 1997 Winter Simulation Conference* (pp. 334–341). Piscataway NJ: IEEE Computer Society.

Chen, J., Chen, C., & Kelton, D. (2003). Optimal computing budget allocation of indifference-zone-selection procedures. Working paper, taken from `http://www.cba.uc.edu/faculty/keltonwd`. Cited 6 January 2005.

Chiarandini, M., Birattari, M., Socha, K., & Rossi-Doria, O. (2003). An effective hybrid approach for the university course timetabling problem. Accepted for publication in the Journal of Scheduling. `http://www.imada.sdu.dk/\~\/marco`. Cited 6 October 2005.

Chiarandini, M. & Stützle, T. (2002). *Experimental Evaluation of Course Timetabling Algorithms.* Technical Report AIDA-02-05, FG Intellektik, TU Darmstadt, Darmstadt, Germany.

Clerc, M. & Kennedy, J. (2002). The particle swarm-explosion, stability, and convergence in a multidimensional complex space. *IEEE Transactions on Evolutionary Computation*, 6(1), 58–73.

Cohen, J. (1990). Things I have learned (so far). *American Psychologist*, 45, 1304–1312.

Cohen, P. R. (1995). *Empirical Methods for Artificial Intelligence.* Cambridge MA: MIT Press.

Cohen, P. R., Gent, I. P., & Walsh, T. (2000). Empirical methods for AI, tutorial given at AAAI, ECAI and Tableaux conferences in 2000. `http://www-users.cs.york.ac.uk/\~\/tw/empirical.html`. Cited 3 January 2004.

Coleman, D. E. & Montgomery, D. C. (1993). A systematic approach to planning for a designed industrial experiment. *Technometrics*, 35, 1–27.

Collett, D. (1991). *Modelling Binary Data.* London: Chapman and Hall.

Cox, D. R. & Hinkley, D. V. (1974). *Theoretical Statistics.* London: Chapman and Hall.

Crites, R. & Barto, A. (1998). Elevator group control using multiple reinforcement learning agents. *Machine Learning*, 33(2-3), 235–262.

Croarkin, C. & Tobias, P., Eds. (2004). *NIST/SEMATECH e-Handbook of Statistical Methods.* National Institute of Standards and Technology. `http://www.itl.nist.gov/div898/handbook`. Cited 15 April 2004.

de Groot, A. (1946/1978). *Thought and Choice in Chess.* New York NY: Mouton.

de Vegt, M. (2005). *Einfluss verschiedener Parametrisierungen auf die Dynamik des Partikel-Schwarm-Verfahrens: Eine empirische Analyse.* Interner Bericht der Systems Analysis Research Group SYS–3/05, Universität Dortmund, Fachbereich Informatik, Germany.

Demetrescu, C. & Italiano, G. F. (2000). What do we learn from experimental algorithmics? In *MFCS '00: Proceedings of the 25th International Symposium on Mathematical Foundations of Computer Science* (pp. 36–51). Berlin, Heidelberg, New York: Springer.

Dolan, E. D. & More, J. J. (2002). Benchmarking optimization software with performance profiles. *Mathematical Programming*, 91, 201–213.

Draper, N. R. & Smith, H. (1998). *Applied Regression Analysis.* New York NY: Wiley, 3rd edition.

Driml, M. & Hanš, O. (1967). On a randomized optimization procedure. In J. Kožešnik (Ed.), *Transactions of the 4th Prague Conference on Information Theory, Statistical Decision Functions and Random Processes* (pp. 273–276). Prague, Czech Republic: Czechoslovak Academy of Sciences.

Droste, S., Jansen, T., & Wegener, I. (2000). *Optimization with Randomized Search Heuristics: The (A)NFL Theorem, Realistic Scenarios, and Difficult Functions.*

Interner Bericht des Sonderforschungsbereichs 531 *Computational Intelligence* CI–91/00, Universität Dortmund, Germany.

Dueck, G. & Scheuer, T. (1990). Threshold accepting: a general purpose optimization algorithm appearing superior to simulated annealing. *Journal of Computational Physics*, 90, 161–175.

Eberhart, R. & Shi, Y. (1998). Comparison between genetic algorithms and particle swarm optimization. In V. Porto, N. Saravanan, D. Waagen, & A. Eiben (Eds.), *Evolutionary Programming*, volume VII (pp. 611–616). Berlin, Heidelberg, New York: Springer.

Efron, B. & Tibshirani, R. J. (1993). *An Introduction to the Bootstrap.* London: Chapman and Hall.

Eiben, A., Hinterding, R., & Michalewicz, Z. (1999). Parameter control in evolutionary algorithms. *IEEE Transactions on Evolutionary Computation*, 3(2), 124–141.

Eiben, A. & Jelasity, M. (2002). A critical note on experimental research methodology in EC. In *Proceedings of the 2002 Congress on Evolutionary Computation (CEC'2002)* (pp. 582–587). Piscataway NJ: IEEE.

Eiben, A. E. & Smith, J. E. (2003). *Introduction to Evolutionary Computing.* Berlin, Heidelberg, New York: Springer.

Emmerich, M., Beume, N., & Naujoks, B. (2005). An EMO algorithm using the hypervolume measure as selection criterion. In C. A. C. Coello, A. H. Aguirre, & E. Zitzler (Eds.), *Proceeding Evolutionary Multi-Criterion Optimization: Third International Conference (EMO 2005)*, volume 3410 of *Lecture Notes in Computer Science* (pp. 62–76). Berlin, Heidelberg, New York: Springer.

Emmerich, M., Giotis, A., Özdemir, M., Bäck, T., & Giannakoglou, K. (2002). Metamodel-assisted evolution strategies. In J. J. M. Guervós, P. Adamidis, H.-G. Beyer, J. L. Fernández-Villacañas, & H.-P. Schwefel (Eds.), *Parallel Problem Solving from Nature—PPSN VII, Proceedings Seventh International Conference, Granada* (pp. 361–370). Berlin, Heidelberg, New York: Springer.

Emmerich, M. & Naujoks, B. (2004). Metamodel-assisted multiobjective optimisation strategies and their application in airfoil design. In I. C. Parmee (Ed.), *Adaptive Computing in Design and Manufacture VI* (pp. 249–260). Berlin, Heidelberg, New York: Springer.

Encyclopaedia Britannica Online (2001). "Faraday effect". `http://members.eb.com/bol/topic?eu=34314\&sctn=1`. Cited 28 October 2001.

Fabian, V. (1962). On multiple decision methods for ranking population means. *Annals of Mathmatical Statistics*, 33, 248–254.

Federov, V. (1972). *Theory of Optimal Experiments.* New York NY: Academic.

Feldt, R. & Nordin, P. (2000). Using factorial experiments to evaluate the effect of genetic programming parameters. In R. Poli & others (Eds.), *Genetic Programming, Proceedings of EuroGP'2000*, volume 1802 of *Lecture Notes in Computer Science* (pp. 271–282). Berlin, Heidelberg, New York: Springer.

Felscher, W. (1998). Two dicta. Historia Matematica Mailing List Archive. `http://archives.math.utk.edu//hypermail/historia/aug98`. Cited 10 May 2004.

Fisher, R. A. (1935). *The Design of Experiments.* Edinburgh: Oliver and Boyd.

Fletcher, R. (1970). A new approach to variable metric algorithms. *Computer Journal*, 13, 317–322.

Folks, J. L. (1981). *Ideas of Statistics.* New York NY: Wiley.

Forster, M. & Sober, E. (1994). How to tell when simpler, more unified, or less ad hoc theories will provide more accurate predictions. *British Journal for the Philosophy of Science*, 45, 1–35.

François, O. & Lavergne, C. (2001). Design of evolutionary algorithms—a statistical perspective. *IEEE Transactions on Evolutionary Computation*, 5(2), 129–148.

Franco, J. & Paull, M. (1983). Probabilistic analysis of the Davis Putnam procedure for solving the satisfiability problem. *Discrete Applied Mathematics*, 5(1), 77–87.

Franklin, A., Ed. (1990). *Experiment, Right or Wrong*. Cambridge, U.K.: Cambridge University Press.

Franklin, A. (2003). Experiment in physics. In E. N. Zalta (Ed.), *The Stanford Encyclopedia of Philosophy*. Stanford CA: Stanford University. `http://plato.stanford.edu/archives/sum2003/entries/physics-experiment`. Cited 14 April 2004.

Galison, P. (1987). *How Experiments End*. Chicago IL: The University of Chicago Press.

Gentle, J. E., Härdle, W., & Mori, Y. (2004a). Computational statistics: An introduction. In J. E. Gentle, W. Härdle, & Y. Mori (Eds.), *Computational Statistics* (pp. 3–16). Berlin, Heidelberg, New York: Springer.

Gentle, J. E., Härdle, W., & Mori, Y., Eds. (2004b). *Handbook of Computational Statistics*. Berlin, Heidelberg, New York: Springer.

Giere, R. N. (1999). Using models to represent reality. In L. Magnani (Ed.), *Model Based Reasoning in Scientific Discovery. Proceedings of the International Conference on Model-Based Reasoning in Scientific Discovery* (pp. 41–57). New York NY: Kluwer.

Gigerenzer, G. (2003). Where do new ideas come from? A heuristic of discovery in cognitive sciences. In M. C. Galavotti (Ed.), *Observation and Experiment in the Natural and Social Sciences* (pp. 99–139). Dordrecht, The Netherlands: Kluwer.

Gigerenzer, G. & Selten, R., Eds. (2002). *Bounded Rationality: The Adaptive Toolbox*. Cambridge MA: MIT Press.

Gigerenzer, G., Todd, P. M., & the ABC research group (1999). *Simple Heuristics That Make Us Smart*. New York NY: Oxford University Press.

Giunta, A., Wojtkiewicz Jr., S., & Eldred, M. (2003). Overview of modern design of experiments methods for computational simulations. In *Proceedings of the 41st AIAA Aerospace Sciences Meeting and Exhibit* Reno NV: American Institute of Aeronautics and Astronautics. Paper AIAA-2003-0649.

Goldberg, A. (1979). *On the Complexity of the Satisfiability Problem*. Technical Report 16, Courant Computer Science Report, New York University, NY.

Goldberg, A., Purdom, P. W., & Brown, C. A. (1982). Average time analyses of simplified Davis-Putnam procedures. *Information Processing Letters*, 15(2), 72–75.

Goldberg, D. E. (1989). *Genetic Algorithms in Search, Optimization, and Machine Learning*. Reading MA: Addison-Wesley.

Goldfarb, D. (1970). A family of variable metric updates derived by variational means. *Mathematics of Computing*, 24, 23–26.

Goldsman, D. & Nelson, B. L. (1998). Statistical screening, selection, and multiple comparison procedures in computer simulation. In D. Medeiros, E. Watson, J. Carson, & M. Manivannan (Eds.), *WSC '98: Proceedings of the 30th Winter Simulation Conference* (pp. 159–166). Los Alamitos CA: IEEE Computer Society.

Goldstein, D., Gigerenzer, G., Hogart, R., Kacelnik, A., Kareev, Y., Klein, G., Martignon, L., Payne, J., & Schlag, K. (2002). Group report: Why and when do simple heuristics work? In G. Gigerenzer & R. Selten (Eds.), *Bounded Rationality: The Adaptive Toolbox* (pp. 174–190). Cambridge MA: MIT Press.

Gooding, D., Pinch, T., & Schaffer, S. (1989). *The Uses of Experiment: Studies in the Natural Sciences.* Cambridge, U.K.: Cambridge University Press.

Gregoire, T. (2001). Biometry in the 21st century: Whither statistical inference? (invited keynote). Proceedings of the Forest Biometry and Information Science Conference held at the University of Greenwich, June 2001. `http://cms1.gre.ac.uk/conferences/iufro/proceedings/gregoire.pdf`. Cited 19 May 2004.

Gregory, D. E., Gao, L., Rosenberg, A. L., & Cohen, P. R. (1996). An empirical study of dynamic scheduling on rings of processors. In *Proceedings of the 8th IEEE Symposium on Parallel and Distributed Processing, SPDP'96 (New Orleans, Louisiana, October 23-26, 1996)* (pp. 470–473). Los Alamitos CA: IEEE Computer Society.

Guala, F. (2003). Experimental localism and external validity. *Philosophy of Science*, 70, 1195–1205.

Gupta, S. S. (1965). On some multiple decision (selection and ranking) rules. *Technometrics*, 7, 225–245.

Hacking, I. (1983). *Representing and Intervening.* Cambridge, U.K.: Cambridge University Press.

Hacking, I. (1996). *Einführung in die Philosophie der Naturwissenschaften.* Stuttgart, Germany: Reclam.

Hacking, I. (2001). *An Introduction to Probability and Inductive Logic.* Cambridge, U.K.: Cambridge University Press.

Hartmann, M. (1988). An improvement on Paulsson's sequential ranking procedure. *Sequential Analysis*, 7, 363–372.

Hartmann, M. (1991). An improvement on Paulsson's procedure for selecting the population with the largest mean from k normal populations with a common unknown variance. *Sequential Analysis*, 10, 1–16.

Hastie, T., Tibshirani, R., & Friedman, J. (2001). *The Elements of Statistical Learning.* Berlin, Heidelberg, New York: Springer.

Hawking, S. W. (1980). *Is the End in Sight for Theoretical Physics?: An Inaugural Lecture.* Cambridge, U.K.: Cambridge University Press.

Hillstrom, K. E. (1977). A simulation test approach to the evaluation of nonlinear optimization algorithms. *ACM Transactions on Mathematical Software*, 3(4), 305–315.

Hooker, J. (1994). Needed: An empirical science of algorithms. *Operations Research*, 42(2), 201–212.

Hooker, J. (1996). Testing heuristics: We have it all wrong. *Journal of Heuristics*, 1(1), 33–42.

Hoos, H. H. (1998). *Stochastic Local Search—Methods, Models, Applications.* PhD thesis, Technische Universität Darmstadt, Germany.

Hoos, H. H. & Stützle, T. (2005). *Stochastic Local Search—Foundations and Applications.* Amsterdam, The Netherlands: Elsevier.

Isaaks, E. H. & Srivastava, R. M. (1989). *An Introduction to Applied Geostatistics.* Oxford, U.K.: Oxford University Press.

Jansen, T. & Wegener, I. (2000). *Evolutionary Algorithms: How to Cope With Plateaus of Constant Fitness and When to Reject Strings of the Same Fit-*

ness. Technical Report CI–96/00, Universität Dortmund, Fachbereich Informatik, Germany.

Jarvie, I. C. (1998). Popper, Karl Raimund. In E. Craig (Ed.), *Routledge Encyclopedia of Philosophy.* London: Routledge. `http://www.rep.routledge.com/article/DD052SECT2`. Cited 19 November 2003.

Jin, R., Chen, W., & Sudjitanto, A. (2002). On sequential sampling for global metamodeling in engineering design. In *Proceedings of the DET02: ASME 2002 Design Engineering Technical Conferences and Computers and Information in Engineering Conference* (pp. 1–10). Montreal, Canada: ASME. DETC2002/DAC-34092.

Jin, Y. (2003). A comprehensive survey of fitness approximation in evolutionary computation. *Soft Computing*, 9(1), 3–12.

Jin, Y. & Branke, J. (2005). Evolutionary optimization in uncertain environments—a survey. *IEEE Transactions on Evolutionary Computation*, 9(3), 303–317.

Johnson, D. S. (2002). A theoretician's guide to the experimental analysis of algorithms. In M. H. Goldwasser, D. S. Johnson, & C. C. McGeoch (Eds.), *Data Structures, Near Neighbor Searches, and Methodology: Fifth and Sixth DIMACS Implementation Challenges* (pp. 215–250). Providence RI: American Mathematical Society.

Johnson, D. S., Aragon, C. R., McGeoch, L. A., & Schevon, C. (1989). Optimization by simulated annealing: an experimental evaluation. Part I, graph partitioning. *Operations Research*, 37(6), 865–892.

Johnson, D. S., Aragon, C. R., McGeoch, L. A., & Schevon, C. (1991). Optimization by simulated annealing: an experimental evaluation. Part II, graph coloring and number partitioning. *Operations Research*, 39(3), 378–406.

Jones, D., Schonlau, M., & Welch, W. (1998). Efficient global optimization of expensive black-box functions. *Journal of Global Optimization*, 13, 455–492.

Kan, A. H. G. R. (1976). *Machine Scheduling Problems: Classification, Complexity and Computation.* The Hague, The Netherlands: Nijhoff.

Kelton, W. (2000). Experimental design for simulation. In J. Joines, R. Barton, K. Kang, & P. Fishwick (Eds.), *Proceedings of the 2000 Winter Simulation Conference* (pp. 32–38). Piscataway NJ: IEEE.

Kempthorne, O. & Folks, L. (1971). *Probability, Statistics, and Data Analysis.* Ames IA: Iowa State University Press.

Kennedy, J. (2003). Bare bones particle swarms. In *Proceedings 2003 IEEE Swarm Intelligence Symposium* (pp. 80–87). Piscataway NJ: IEEE.

Kennedy, J. & Eberhart, R. (1995). Particle swarm optimization. In *Proceedings IEEE International Conference on Neural Networks*, volume IV (pp. 1942–1948). Piscataway NJ: IEEE.

Kennedy, J. & Eberhart, R. (2001). *Swarm Intelligence.* San Francisco CA: Morgan Kaufmann.

Kieseppä, I. A. (1997). Akaike information criterion, curve-fitting and the philosophical problem of simplicity. *British Journal for the Philosophy of Science*, 48, 21–48.

Kim, S.-H. & Nelson, B. L. (2001). A fully sequential procedure for indifference-zone selection in simulation. *ACM Transactions on Modeling and Computer Simulation*, 11(3), 251–273.

Kleijnen, J. P. C. (1987). *Statistical Tools for Simulation Practitioners.* New York NY: Marcel Dekker.

Kleijnen, J. P. C. (1997). Experimental design for sensitivity analysis, optimization, and validation of simulation models. In J. Banks (Ed.), *Handbook of Simulation*. New York NY: Wiley.

Kleijnen, J. P. C. (2001). *Experimental Design for Sensitivity Analysis of Simulation Models*. Discussion Paper 15, Center for Economic Research, Tilburg University, The Netherlands.

Kleijnen, J. P. C. & Van Groenendaal, W. (1992). *Simulation—A Statistical Perspective*. Chichester, U.K.: Wiley.

Klein, G. (2002). The fiction of optimization. In G. Gigerenzer & R. Selten (Eds.), *Bounded Rationality: The Adaptive Toolbox* (pp. 103–121). Cambridge MA: MIT Press.

Knuth, D. (1981). *The Art of Computer Programming*. Reading MA: Addison-Wesley, 2nd edition.

Kursawe, F. (1999). *Grundlegende empirische Untersuchungen der Parameter von Evolutionsstrategien – Metastrategien*. Dissertation, Fachbereich Informatik, Universität Dortmund, Germany.

Lagarias, J. C., Reeds, J. A., Wright, M. H., & Wright, P. E. (1998). Convergence properties of the Nelder–Mead simplex method in low dimensions. *SIAM Journal on Optimization*, 9(1), 112–147.

Lasarczyk, C. W. G. & Banzhaf, W. (2005a). An algorithmic chemistry for genetic programming. In M. Keijzer, A. Tettamanzi, P. Collet, J. I. van Hemert, & M. Tomassini (Eds.), *Proceedings of the 8th European Conference on Genetic Programming*, volume 3447 of *Lecture Notes in Computer Science* (pp. 1–12). Berlin, Heidelberg, New York: Springer.

Lasarczyk, C. W. G. & Banzhaf, W. (2005b). Total synthesis of algorithmic chemistries. In H.-G. Beyer & others (Eds.), *Proceedings Genetic and Evolutionary Computation Conference (GECCO 2005), Washington D.C.*, volume 2 (pp. 1635–1640). New York NY: Association for Computing Machinery.

Law, A. & Kelton, W. (2000). *Simulation Modeling and Analysis*. New York NY: McGraw-Hill, 3rd edition.

Lewis, R., Torczon, V., & Trosset, M. (2000). Direct search methods: Then and now. *Journal of Computational and Applied Mathematics*, 124(1–2), 191–207.

Lophaven, S., Nielsen, H., & Søndergaard, J. (2002a). *Aspects of the Matlab Toolbox DACE*. Technical Report IMM-REP-2002-13, Informatics and Mathematical Modelling, Technical University of Denmark, Copenhagen, Denmark.

Lophaven, S., Nielsen, H., & Søndergaard, J. (2002b). *DACE—A Matlab Kriging Toolbox*. Technical Report IMM-REP-2002-12, Informatics and Mathematical Modelling, Technical University of Denmark, Copenhagen, Denmark.

Mammen, E. & Nandi, S. (2004). Bootstrap and resampling. In J. E. Gentle, W. Härdle, & Y. Mori (Eds.), *Handbook of Computational Statistics* (pp. 467–495). Berlin, Heidelberg, New York: Springer.

Markon, S. (1995). *Studies on Applications of Neural Networks in the Elevator System*. PhD thesis, Kyoto University, Japan.

Markon, S., Arnold, D. V., Bäck, T., Beielstein, T., & Beyer, H.-G. (2001). Thresholding—A selection operator for noisy ES. In J.-H. Kim, B.-T. Zhang, G. Fogel, & I. Kuscu (Eds.), *Proceedings 2001 Congress on Evolutionary Computation (CEC'01), Seoul* (pp. 465–472). Piscataway NJ: IEEE.

Markon, S., Kita, H., Kise, H., & Bartz-Beielstein, T., Eds. (2006). *Modern Supervisory and Optimal Control with Applications in the Control of Passenger Traffic Systems in Buildings.* Berlin, Heidelberg, New York: Springer.

Martinez, W. L. & Martinez, A. R. (2002). *Computational Statistics Handbook with MATLAB.* Boca Raton FL: Chapman & Hall/CRC.

Matyáš, J. (1965). Random Optimization. *Automation and Remote Control*, 26(2), 244–251.

Mayo, D. G. (1983). An objective theory of statistical testing. *Synthese*, 57, 297–340.

Mayo, D. G. (1996). *Error and the Growth of Experimental Knowledge.* Chicago IL: The University of Chicago Press.

Mayo, D. G. (1997). Severe tests, arguing from error, and methodological underdetermination. *Philosophical Studies*, 86, 243–266.

McCullagh, P. & Nelder, J. (1989). *Generalized Linear Models.* London, U.K.: Chapman and Hall, 2nd edition.

McGeoch, C. C. (1986). *Experimental Analysis of Algorithms.* PhD thesis, Carnegie Mellon University, Pittsburgh PA.

McKay, M. D., Beckman, R. J., & Conover, W. J. (1979). A comparison of three methods for selecting values of input variables in the analysis of output from a computer code. *Technometrics*, 21(2), 239–245.

Mehnen, J., Michelitsch, T., Bartz-Beielstein, T., & Henkenjohann, N. (2004a). Systematic analyses of multi-objective evolutionary algorithms applied to real-world problems using statistical design of experiments. In R. Teti (Ed.), *Proceedings Fourth International Seminar Intelligent Computation in Manufacturing Engineering (CIRP ICME'04)*, volume 4 (pp. 171–178). Naples, Italy.

Mehnen, J., Michelitsch, T., Bartz-Beielstein, T., & Lasarczyk, C. W. G. (2005). Multiobjective evolutionary design of mold temperature control using DACE for parameter optimization. In H. Pfützner & E. Leiss (Eds.), *Proceedings Twelfth International Symposium Interdisciplinary Electromagnetics, Mechanics, and Biomedical Problems (ISEM 2005)*, volume L11-1 (pp. 464–465). Vienna, Austria: Vienna Magnetics Group Reports.

Mehnen, J., Michelitsch, T., Bartz-Beielstein, T., & Schmitt, K. (2004b). Evolutionary optimization of mould temperature control strategies: Encoding and solving the multiobjective problem with standard evolution strategy and kit for evolutionary algorithms. *Journal of Engineering Manufacture (JEM)*, 218(B6), 657–665.

Merriam-Webster Online Dictionary (2004). "Theory". `http://www.merriam-webster.com`. Cited 2 April 2004.

Mertens, H. (1990). *Moderne – Sprache – Mathematik: eine Geschichte des Streits um die Grundlagen der Disziplin und des Subjekts formaler Systeme.* Frankfurt am Main, Germany: Suhrkamp.

Metropolis, N. & Ulam, S. (1949). The Monte Carlo Method. *Journal of the American Statistical Association*, 44(247), 335–341.

Minsky, M. (1985). *The Society of Mind.* New York NY: Simon and Schuster.

Mitchell, D. G., Selman, B., & Levesque, H. J. (1992). Hard and easy distributions for SAT problems. In P. Rosenbloom & P. Szolovits (Eds.), *Proceedings of the Tenth National Conference on Artificial Intelligence* (pp. 459–465). Menlo Park CA: AAAI.

Montgomery, D. C. (2001). *Design and Analysis of Experiments.* New York NY: Wiley, 5th edition.

More, J. J., Garbow, B. S., & Hillstrom, K. E. (1981). Testing unconstrained optimization software. *ACM Transactions on Mathematical Software*, 7(1), 17–41.

Moret, B. M. E. (2002). Towards a discipline of experimental algorithmics. In M. Goldwasser, D. Johnson, & C. McGeoch (Eds.), *Data Structures, Near Neighbor Searches, and Methodology: Fifth and Sixth DIMACS Implementation Challenges, DIMACS Monographs 59* (pp. 197–213). Providence RI: American Mathematical Society.

Morgan, J. & Sonquist, J. (1963). Problems in the analysis of survey data and a proposal. *Journal of the American Statistical Association*, 58, 415–434.

Morrison, D. E. & Henkel, R. E., Eds. (1970). *The Significance Test Controversy—A Reader*. London, U.K.: Butterworths.

Mott, N. F. & Sneddon, I. N. (1948). *Wave Mechanics and Its Application*. London, U.K.: Oxford University Press.

Myers, R. & Hancock, E. (2001). Empirical modelling of genetic algorithms. *Evolutionary Computation*, 9(4), 461–493.

Nagylaki, T. (1992). *Introduction to Theoretical Population Genetics*. Berlin, Heidelberg, New York: Springer.

Naudts, B. & Kallel, L. (2000). A comparison of predictive measures of problem difficulty in evolutionary algorithms. *IEEE Transactions on Evolutionary Computation*, 4(1), 1–15.

Naujoks, B., Beume, N., & Emmerich, M. (2005a). Metamodel-assisted SMS-EMOA applied to airfoil optimization tasks. In R. Schilling, W. Haase, J. Périaux, & H. Baier (Eds.), *Proceedings EUROGEN'05 (CD-ROM)*. München, Germany: Technische Universität.

Naujoks, B., Beume, N., & Emmerich, M. (2005b). Multi-objective optimisation using S-metric selection: Application to three dimensional solution spaces. In B. McKay & others (Eds.), *Proceedings 2005 Congress on Evolutionary Computation*, volume 2 (pp. 1282–1289). Piscataway NJ: IEEE.

Nelder, J. & Mead, R. (1965). A simplex method for function minimization. *Computer Journal*, 7, 308–313.

Nelson, B., Swann, J., Goldsman, D., & Song, W. (1998). *Simple Procedures for Selecting the Best Simulated System When the Number of Alternatives Is Large*. Technical report, Dept. of Industrial Engineering and Management Science, Northwestern University, Evanston, Illinois.

Neumaier, A., Shcherbina, O., Huyer, W., & Vinko, T. (2005). A comparison of complete global optimization solvers. *Mathematical Programming B*, 103, 335–356.

Newman, J. R., Ed. (1956). *The World of Mathematics*. New York NY: Simon and Schuster.

Neyman, J. (1950). *First Course in Probability and Statistics*. New York NY: Henry Holt.

Niedermeier, R. (2003). Parametrisierte Algorithmen. Lecture Notes, Universität Tübingen, Germany. `http://www-fs.informatik.uni-tuebingen.de/lehre/ws02-03/paramalg.htm`. Cited 20 December 2004.

Nocedal, J. & Wright, S. (1999). *Numerical Optimization*. Berlin, Heidelberg, New York: Springer.

Parkes, A. J. & Walser, J. P. (1996). Tuning local search for satisfiability testing. In *Proceedings of the Thirteenth National Conference on Artificial Intelligence (AAAI'96)* (pp. 356–362).

Parsopoulos, K. & Vrahatis, M. (2002). Recent approaches to global optimization problems through particle swarm optimization. *Natural Computing*, 1(2–3), 235–306.

Parsopoulos, K. E. & Vrahatis, M. N. (2004). On the computation of all global minimizers through particle swarm optimization. *IEEE Transactions on Evolutionary Computation*, 8(3), 211–224.

Paulson, E. (1964). A sequential procedure for selecting the population with the largest mean from k normal populations. *Annals of Mathematical Statistics*, 35, 174–180.

Pichitlamken, J. & Nelson, B. L. (2001). Comparing systems via stochastic simulation: Selection-of-the-best procedures for optimization via simulation. In *Proceedings of the 33rd Winter Simulation Conference* (pp. 401–407). Washington DC: IEEE Computer Society.

Pichitlamken, J., Nelson, B. L., & Hong, L. J. (2003). A sequential procedure for neighborhood selection-of-the-best in optimization via simulation. Working Paper. `http://www.ielm.ust.hk/dfaculty/hongl/`. Cited 18 June 2004.

Popper, K. (1959). *The Logic of Scientific Discovery.* London, U.K.: Hutchinson.

Popper, K. (1979). *Objective Knowledge: An Evolutionary Approach.* Oxford, U.K.: Oxford University Press.

Popper, K. (1983). *Realisim and the Aim of Science.* Totowa NJ: Rowman and Littlefield.

Press, W. H., Teukolsky, S. A., Vetterling, W. T., & Flannery, B. P. (1992). *Numerical Recipes in Fortran 77.* Cambridge, U.K.: Cambridge University Press.

Preuß, M. & Bartz-Beielstein, T. (2006). Self-adaptation in evolution strategies—an experimental analysis based on sequential parameter optimization. In F. Lobo, C. Lima, & Z. Michalewicz (Eds.), *Parameter Setting in Evolutionary Algorithms*, Studies in Computational Intelligence. Berlin, Heidelberg, New York: Springer.

Pukelsheim, F. (1993). *Optimal Design of Experiments.* New York NY: Wiley.

Rardin, R. & Uzsoy, R. (2001). Experimental evaluation of heuristic optimization algorithms: A tutorial. *Journal of Heuristics*, 7(3), 261–304.

Rechenberg, I. (1973). *Evolutionsstrategie. Optimierung technischer Systeme nach Prinzipien der biologischen Evolution.* Stuttgart, Germany: frommann–holzboog.

Reeves, C. & Yamada, T. (1998). Genetic algorithms, path relinking and the flowshop sequencing problem. *Evolutionary Computation Journal*, 6(1), 230–234.

Rosenbrock, H. (1960). An automatic method for finding the greatest or least value of a function. *Computer Journal*, 3, 175–184.

Roth, A. J. (1978). A new procedure for selecting a subset containing the best normal population. *Journal American Statistical Association*, 73, 613–617.

Rubin, H. (1971). Occam's razor needs new blades. In V. Godambe & D. Sprott (Eds.), *Foundations of Statistical Inference* (pp. 372–374). Toronto, Canada: Holt, Rinehart and Winston.

Rubinstein, A. (1998). *Modeling Bounded Rationality.* Cambridge MA: MIT Press.

Rudolph, G. (1997a). *Convergence Properties of Evolutionary Algorithms.* Hamburg, Germany: Kovač.

Rudolph, G. (1997b). Reflections on bandit problems and selection methods in uncertain environments. In T. Bäck (Ed.), *Genetic Algorithms: Proceedings Seventh*

International Conference (ICGA'97) (pp. 166–173). San Francisco CA: Morgan Kaufmann.

Sacks, J., Welch, W. J., Mitchell, T. J., & Wynn, H. P. (1989). Design and analysis of computer experiments. *Statistical Science*, 4(4), 409–435.

Sanders, P. (2004). Announcement of the *Algorithm Engineering for Fundamental Data Structures and Algorithms* talk during the Summer School on Experimental Algorithmics. `http://www.diku.dk/forskning/performance-engineering/Sommerskole/scientific-program.html`. Cited 17 May 2004.

Sano, Y. & Kita, H. (2000). Optimization of noisy fitness functions by means of genetic algorithms using history of search. In M. Schoenauer & others (Eds.), *Parallel Problem Solving from Nature (PPSN VI)*, volume 1917 of *Lecture Notes in Computer Science* (pp. 571–580). Berlin, Heidelberg, New York: Springer.

Santner, T. J. (1976). A two-stage procedure for selection of δ^*-optimal means in the normal case. *Communications in Statistics—Theory and Methods*, A5, 283–292.

Santner, T. J., Williams, B. J., & Notz, W. I. (2003). *The Design and Analysis of Computer Experiments.* Berlin, Heidelberg, New York: Springer.

Satterthwaite, F. E. (1959a). Random balance experimentation. *Technometrics*, 1, 111–137.

Satterthwaite, F. E. (1959b). *REVOP or Random Evolutionary Operation.* Technical Report Report 10-10-59, Merrimack College, North Andover MA.

Schaffer, J. D., Caruana, R. A., Eshelman, L., & Das, R. (1989). A study of control parameters affecting online performance of genetic algorithms for function optimization. In J. D. Schaffer (Ed.), *Proceedings of the Third International Conference on Genetic Algorithms* (pp. 51–60). San Mateo CA: Morgan Kaufman.

Schmidt, J. W. (1986). Introduction to systems analysis, modeling and simulation. In J. Wilson, J. Henriksen, & S. Roberts (Eds.), *WSC '86: Proceedings of the 18th Winter Simulation Conference* (pp. 5–16). New York NY: Association for Computing Machinery.

Schneier, B. (1996). *Applied Cryptography: Protocols, Algorithms, and Source Code in C.* New York NY: Wiley.

Schonlau, M. (1997). *Computer Experiments and Global Optimization.* PhD thesis, University of Waterloo, Ontario, Canada.

Schwefel, H.-P. (1975). *Evolutionsstrategie und numerische Optimierung.* Dr.-Ing. Dissertation, Technische Universität Berlin, Fachbereich Verfahrenstechnik, Berlin, Germany.

Schwefel, H.-P. (1977). *Numerische Optimierung von Computer–Modellen mittels der Evolutionsstrategie*, volume 26 of *Interdisciplinary Systems Research.* Basel, Switzerland: Birkhäuser.

Schwefel, H.-P. (1979). Direct search for optimal parameters within simulation models. In R. D. Conine, E. D. Katz, & J. E. Melde (Eds.), *Proceedings Twelfth Annual Simulation Symposium, Tampa FL* (pp. 91–102). Long Beach CA: IEEE Computer Society.

Schwefel, H.-P. (1981). *Numerical Optimization of Computer Models.* Chichester, U.K.: Wiley.

Schwefel, H.-P. (1988). Evolutionary learning optimum-seeking on parallel computer architectures. In A. Sydow, S. G. Tzafestas, & R. Vichnevetsky (Eds.), *Systems Analysis and Simulation*, volume 1 (pp. 217–225). Berlin, Germany: Akademie.

Schwefel, H.-P. (1995). *Evolution and Optimum Seeking.* Sixth-Generation Computer Technology. New York NY: Wiley.

Schwefel, H.-P., Rudolph, G., & Bäck, T. (1995). *Contemporary evolution strategies.* Interner Bericht der Systems Analysis Research Group SYS–6/95, Fachbereich Informatik, Universität Dortmund, Germany.

Schwefel, H.-P., Wegener, I., & Weinert, K., Eds. (2003). *Advances in Computational Intelligence—Theory and Practice.* Berlin, Heidelberg, New York: Springer.

Selvin, H. C. (1970). A critique of tests of significance in survey research. In D. Morrison & R. Henkel (Eds.), *The Significance Test Controversy—A Reader* (pp. 94–106). London, U.K.: Butterworths.

Shanno, D. F. (1970). Conditioning of quasi-Newton methods for function minimization. *Mathematics of Computing*, 24, 647–656.

Shi, Y. (2004). Particle swarm optimization. *IEEE CoNNectionS – The Newsletter of the IEEE Neural Networks Society*, 2(1), 8–13.

Shi, Y. & Eberhart, R. (1999). Empirical study of particle swarm optimization. In P. J. Angeline, Z. Michalewicz, M. Schoenauer, X. Yao, & A. Zalzala (Eds.), *Proceedings of the Congress of Evolutionary Computation*, volume 3 (pp. 1945–1950). Piscataway NJ: IEEE.

Simon, H. (1955). A behavioral model of rational choice. *Quarterly Journal of Economics*, 69(1), 99–118.

Simpson, T. W., Booker, A., Ghosh, D., Giunta, A. A., Koch, P., & Yang, R.-J. (2004). Approximation methods in multidisciplinary analysis and optimization: a panel discussion. *Structural and Multidisciplinary Optimization*, 27, 302–313.

Singer, S. & Singer, S. (2004). Efficient termination test for the Nelder-Mead search algorithm. In T. Simos & C. Tsitouras (Eds.), *International Conference on Numerical Analysis and Applied Mathematics 2004 (ICNAAM)* (pp. 348–351). Weinheim, Germany: Wiley.

Smith, V. (1962). An experimental study of competitive market behavior. *Journal of Political Economy*, 70, 111–137.

So, A. & Chan, W. (1999). *Intelligent Building Systems.* Dordrecht, The Netherlands: Kluwer.

Spall, J. (2003). *Introduction to Stochastic Search and Optimization.* Hoboken, NJ: Wiley.

Stagge, P. (1998). Averaging efficiently in the presence of noise. In A. Eiben (Ed.), *Parallel Problem Solving from Nature, PPSN V* (pp. 188–197). Berlin, Heidelberg, New York: Springer.

Staley, K. (2002). What experiment did we just do? Counterfactual error statistics and uncertainties about the reference class. *Philosophy of Science*, 69, 279–299.

Stewart, E. C., Kavanaugh, W. P., & Brocker, D. H. (1967). Study of a global search algorithm for optimal control. In *Proceedings of the 5th International Analogue Computation Meeting, Lausanne* (pp. 207–230). Brussels, Belgium: Presses Academiques Europeennes.

Suganthan, P. N., Hansen, N., Liang, J. J., Deb, K., Chen, Y.-P., Auger, A., & Tiwari, S. (2005). *Problem Definitions and Evaluation Criteria for the CEC 2005 Special Session on Real-Parameter Optimization.* Technical report, Nanyang Technological University, Singapore.

Sullivan, D. W. & Wilson, J. R. (1984). Restricted subset selection for normal populations with unknown and unequal variances. In *Proceedings of the 1984 Winter Simulation Conference* (pp. 266–274). Dallas TX.

Sullivan, D. W. & Wilson, J. R. (1989). Restricted subset selection procedures for simulation. *Operations Research*, 61, 585–592.

Suppes, P. (1969a). A comparison of the meaning and uses of models in mathematics and the empirical sciences. In P. Suppes (Ed.), *Studies in the Methodology and Foundation of Science* (pp. 11–13). Dordrecht, The Netherlands: Reidel.

Suppes, P. (1969b). Models of data. In P. Suppes (Ed.), *Studies in the Methodology and Foundation of Science* (pp. 24–35). Dordrecht, The Netherlands: Reidel.

Suppes, P. (1969c). *Studies in the Methodology and Foundation of Science.* Dordrecht, The Netherlands: Reidel.

Tarski, A. (1953). A general method in proofs of undecidability. In A. Tarski, A. Mostowski, & R. M. Robinson (Eds.), *Undecidable Theories* (pp. 3–35). Amsterdam, The Netherlands: North-Holland.

Tarski, A., Mostowski, A., & Robinson, R. M., Eds. (1953). *Undecidable Theories.* Amsterdam, The Netherlands: North-Holland.

Therneau, T. M. & Atkinson, E. J. (1997). *An Introduction to Recursive Partitioning Using the RPART Routines.* Technical Report 61, Department of Health Science Research, Mayo Clinic, Rochester NY.

Tosic, M. (2006). Evolutionäre Kreuzungsminimierung. Diploma thesis, University of Dortmund, Germany.

Trosset, M. & Padula, A. (2000). *Designing and Analyzing Computational Experiments for Global Optimization.* Technical Report 00-25, Department of Computational and Applied Mathematics, Rice University, Houston TX.

Tukey, J. (1991). The philosophy of multiple comparisons. *Statistical Science*, 6, 100–116.

Van Breedam, A. (1995). Improvement heuristics for the vehicle routing problem based on simulated annealing. *European Journal of Operational Research*, 86, 480–490.

van der Laan, P. (1992). Subset selection of an almost best treatment. *Biometrical Journal*, 34, 647–656.

Watson, J.-P., Barbulescu, L., Howe, A. E., & Whitley, D. (1999). Algorithm performance and problem structure for flow-shop scheduling. In *Proceedings of the Sixteenth National Conference on Artificial Intelligence (AAAI-99)* (pp. 688–695). Cambridge MA: MIT Press.

Weihe, K., Brandes, U., Liebers, A., Müller-Hannemann, M., Wagner, D., & Willhalm, T. (1999). Empirical design of geometric algorithms. In *SCG '99: Proceedings of the Fifteenth Annual Symposium on Computational Geometry* (pp. 86–94). New York NY: Association for Computing Machinery.

Weinert, K., Mehnen, J., Michelitsch, T., Schmitt, K., & Bartz-Beielstein, T. (2004). A multiobjective approach to optimize temperature control systems of moulding tools. *Production Engineering Research and Development, Annals of the German Academic Society for Production Engineering*, XI(1), 77–80.

Welch, W. J., Buck, R. J., Sacks, J., Wynn, H. P., Mitchell, T. J., & Morris, M. D. (1992). Screening, predicting, and computer experiments. *Technometrics*, 34, 15–25.

Whitley, D., Mathias, K., Rana, S., & Dzubera, J. (1996). Evaluating evolutionary algorithms. *Artificial Intelligence*, 85(1–2), 245–276.

Whitley, D., Watson, J., Howe, A., & Barbulescu, L. (2002). *Testing, Evaluation, and Performance of Optimization and Learning Systems.* Technical report, The

GENITOR Research Group in Genetic Algorithms and Evolutionary Computation, Colorado State University, Fort Collins CO.
Whitley, D. L., Mathias, K. E., Rana, S., & Dzubera, J. (1995). Building better test functions. In L. Eshelman (Ed.), *Proceedings of the Sixth International Conference on Genetic Algorithms* (pp. 239–246). San Francisco CA: Morgan Kaufmann.
Wineberg, M. & Christensen, S. (2004). An Introduction to Statistics for EC Experimental Analysis. CEC tutorial slides. `http://ls11-www.cs.uni-dortmund.de/people/tom/public\_html/experiment04.html`. Cited 15 October 2004.
Winker, P. (2001). *Optimization Heuristics in Econometrics: Applications of Threshold Accepting.* Chichester, U.K.: Wiley.
Zoubir, A. M. & Boashash, B. (1998). The bootstrap and its application in signal processing. *Signal Processing Magazine, IEEE*, 15(1), 56–67.

Index

Nomenclature

Roman symbols

#	number sign, p. 113
$\mathbf{1}$	vector of ones, p. 50
$B(x_0, \epsilon)$	ϵ-environment of x_0, p. 70
c_i	passenger waiting bit, p. 72
d	difference vector of two random samples, p. 45
e	experimental outcome, p. 28
$E(X)$	expectation of the random variable X, p. 49
$ET(\cdot)$	experimental testing model, p. 23
f^*	known best objective function value, p. 110
g	generation, p. 97
I_n	n-dimensional identity matrix, p. 50
k	number of design variables (factors), p. 49
l	leaf of a tree, p. 56
$M(\cdot)$	probability model, p. 23
N	total number of observations, p. 49
n	sample size, p. 14
n_b	number of bootstrap samples, p. 45
n_L	the number of leaves in regression tree, p. 57
n_v	number of cases in node v, p. 57
P	probability distribution, p. 23
p	passenger arrival rate, p. 72
q	number of regression parameters, p. 49
r	sample size for reevaluation, p. 146
$R(T)$	mean squared error for the tree T, p. 57
$R(v)$	squared error of the node v, p. 57
$R_{\text{CV}}(\cdot)$	cross-validation estimate for the prediction error, p. 57
$R_{c_p}(\cdot)$	cost-complexity measure, p. 57
$RU(\cdot)$	testing rule, p. 23
s	number of candidates (solutions), p. 146
S^2	sample variance, p. 43
s_ν^2	estimate of σ_ϵ^2 based on ν degrees of freedom, p. 149

s_a	step-size adjustment factor, p. 95
S_d	sample standard deviation of the differences, p. 44
s_i	server present bit, p. 72
s_n	length of the success vector, p. 95
S_p^2	pooled variance, p. 43
s_R	estimate of the standard error of the prediction error, p. 57
s_r	1/success rate, p. 95
s_u	step-size adaptation interval, p. 95
SS_{E}	sum of squares due to error, p. 49
SS_{TREAT}	sum of squares due to the treatments, p. 49
SS_{T}	total corrected sum of squares, p. 49
T	test statistic, p. 23
T_L	set of all leaves of a tree T, p. 56
$t_{\alpha,n}$	upper α percentage point of the t-distribution with n d.f., p. 43
$U[0,1]$	uniformly distributed r.v. from $[0,1]$
v	node of a tree, p. 56
$v^{(t)}$	memory vector, p. 155
v_L	left subtree with root node v, p. 57
v_R	right subtree with root node v, p. 57
X	regression matrix, p. 49
x	input or design variable, p. 48
$x(t)$	state of the system at time t, p. 73
$X^{(0)}$	set of search points at generation 0, p. 88
$x^{(0)}$	starting point, p. 67
x^*	minimizer (global or local), p. 66
X_A	algorithm design, p. 80
x_l	lower initialization bound
X_P	problem design, p. 81
x_u	upper initialization bound
x^*_{ap}	apparent global optimizer, p. 68
x_{border}	border for successful solutions, p. 110
x_{effect}	effects of a subset, p. 61
Y	random variable, p. 23
$Z(\cdot)$	random process, p. 60
$Z_{s,\rho}^{(\alpha)}$	upper α equicoordinate critical point, p. 43
z_α	upper α percentage point of the normal distribution, p. 26
$\mathcal{C}$	scientific claim, p. 23
$\mathcal{D}$	set of all experimental designs, p. 81
$\mathcal{D}_A$	set of all algorithm designs, p. 80
$\mathcal{D}_P$	set of all problem designs, p. 81
$\mathcal{F}$	regression model, p. 59
$\mathcal{R}$	correlation model, p. 60
$\mathcal{O}$	observation, p. 23
$\mathcal{Y}$	sample space, p. 23

Acronyms

AID	automatic interaction detection, p. 55
ANOVA	analysis of variance, p. 48
CART	classification and regression trees, p. 55
CCD	central composite designs, p. 82
CDF	cumulative distribution function, p. 116
CEC	Congress on Evolutionary Computation, p. 4
CI	computational intelligence, p. 70
CR	critical region, p. 24
CRN	common random numbers, p. 19
CS	correct selection, p. 147
DACE	design and analysis of computer experiments, p. 7
DETEQ	deterministically determined starting vectors, p. 88
DETMOD	deterministically modified starting vectors, p. 88
DOE	design of experiments, p. 5
EA	evolutionary algorithm, p. 3
EC	evolutionary computation, p. 3
ES	evolution strategy, p. 61
ESGC	elevator supervisory group control, p. 70
EVOP	evolutionary operation, p. 86
EXH	resources exhausted, p. 89
EXP	exponential correlation function, p. 60
EXPG	general exponential correlation function, p. 60
EXPIMP	expected improvement heuristic, p. 87
FSOL	problem was solved (function values), p. 89
GAUSS	Gaussian correlation function, p. 60
GECCO	Genetic and Evolutionary Computation Conference, p. 4
GL	guidelines, p. 18
GLM	generalized linear model, p. 5
IQR	interquartile range, p. 53
LHD	Latin hypercube design, p. 85
LHS	Latin hypercube sampling, p. 85
MBST	mean best function value, p. 111
MC	Monte Carlo sampling, p. 85
MI	misconstrual, p. 29
MSE	mean squared error of the predictor, p. 60
MTER	Hillstrom's efficiency measure, p. 113
NFL	no free lunch theorem, p. 67
NMS	Nelder–Mead simplex algorithm, p. 93
NN	neural network, p. 70
NP	nondeterministic polynomial, p. 7
NPT	Neyman–Pearson theory of testing, p. 10
NPT*	Mayo's extension of NPT, p. 28

NUNIRND	nonuniform random starts, p. 88
OCBA	optimal computing budget allocation, p. 153
OSL	observed significance level, p. 30
PM	performance measure, p. 106
P	polynomial, p. 7
PRATE	progress rate, p. 115
PSO	particle swarm optimization, p. 13
REVOP	random evolutionary operation, p. 86
RG	research goal, p. 17
RLD	run length distribution, p. 112
RSM	response surface methodology, p. 120
SAT	satisfiability problem, p. 66
SCR	success ratio, p. 50
SPO	sequential parameter optimization, p. 126
SQ	severity question, p. 28
SR	severity requirement, p. 28
ST	statistical test, p. 25
STAL	algorithm stalled, p. 89
TA	threshold acceptance, p. 150
TC	tree construction, p. 56
TR	threshold rejection, p. 150
TS	threshold selection, p. 52
TSP	traveling salesperson problem, p. 7
UNIRND	uniform random starts, p. 88
VRT	variance-reduction techniques, p. 19
XSOL	problem was solved, p. 89

Greek symbols

$\alpha_{\overline{d}}(\delta)$	observed significance level (rejection), p. 30
$\beta_{\overline{d}}(\delta)$	observed significance level (acceptance), p. 33
δ	difference between two population means, p. 25
δ^*	the smallest difference worth detecting, p. 147
δ_{un}	the largest scientifically unimportant value, p. 31
ϵ	machine precision, p. 110
μ	mean of Y, p. 23
$\mu_{[i]}$	ith ordered mean, p. 147
Ω	parameter space, p. 23
π	policy, p. 71
π^*	optimal policy, p. 74
ρ	correlation, p. 43
σ	standard deviation, p. 25
$\sigma^{(0)}$	starting value for the step size, p. 95
$\sigma_{\overline{d}}$	standard error, p. 26
τ	threshold, p. 150
φ	progress rate, p. 115

Natural Computing Series

W.M. Spears: **Evolutionary Algorithms. The Role of Mutation and Recombination.** XIV, 222 pages, 55 figs., 23 tables. 2000

H.-G. Beyer: **The Theory of Evolution Strategies.** XIX, 380 pages, 52 figs., 9 tables. 2001

L. Kallel, B. Naudts, A. Rogers (Eds.): **Theoretical Aspects of Evolutionary Computing.** X, 497 pages. 2001

G. Păun: **Membrane Computing. An Introduction.** XI, 429 pages, 37 figs., 5 tables. 2002

A.A. Freitas: **Data Mining and Knowledge Discovery with Evolutionary Algorithms.** XIV, 264 pages, 74 figs., 10 tables. 2002

H.-P. Schwefel, I. Wegener, K. Weinert (Eds.): **Advances in Computational Intelligence. Theory and Practice.** VIII, 325 pages. 2003

A. Ghosh, S. Tsutsui (Eds.): **Advances in Evolutionary Computing. Theory and Applications.** XVI, 1006 pages. 2003

L.F. Landweber, E. Winfree (Eds.): **Evolution as Computation.** DIMACS Workshop, Princeton, January 1999. XV, 332 pages. 2002

M. Hirvensalo: **Quantum Computing.** 2nd ed., XI, 214 pages. 2004 (first edition published in the series)

A.E. Eiben, J.E. Smith: **Introduction to Evolutionary Computing.** XV, 299 pages. 2003

A. Ehrenfeucht, T. Harju, I. Petre, D.M. Prescott, G. Rozenberg: **Computation in Living Cells. Gene Assembly in Ciliates.** XIV, 202 pages. 2004

L. Sekanina: **Evolvable Components. From Theory to Hardware Implementations.** XVI, 194 pages. 2004

G. Ciobanu, G. Rozenberg (Eds.): **Modelling in Molecular Biology.** X, 310 pages. 2004

R.W. Morrison: **Designing Evolutionary Algorithms for Dynamic Environments.** XII, 148 pages, 78 figs. 2004

R. Paton†, H. Bolouri, M. Holcombe, J.H. Parish, R. Tateson (Eds.): **Computation in Cells and Tissues. Perspectives and Tools of Thought.** XIV, 358 pages, 134 figs. 2004

M. Amos: **Theoretical and Experimental DNA Computation.** XIV, 170 pages, 78 figs. 2005

M. Tomassini: **Spatially Structured Evolutionary Algorithms.** XIV, 192 pages, 91 figs., 21 tables. 2005

G. Ciobanu, G. Păun, M.J. Pérez-Jiménez (Eds.): **Applications of Membrane Computing.** X, 441 pages, 99 figs., 24 tables. 2006

K.V. Price, R.M. Storn, J.A. Lampinen: **Differential Evolution.** XX, 538 pages, 292 figs., 48 tables and CD-ROM. 2006

J. Chen, N. Jonoska, G. Rozenberg: **Nanotechnology: Science and Computation.** XII, 385 pages, 126 figs., 10 tables. 2006

A. Brabazon, M. O'Neill: **Biologically Inspired Algorithms for Financial Modelling.** XVI, 275 pages, 92 figs., 39 tables. 2006

T. Bartz-Beielstein: **Experimental Research in Evolutionary Computation.** XIV, 214 pages, 66 figs., 36 tables. 2006

Printing: Krips bv, Meppel
Binding: Stürtz, Würzburg